THE MIDDLE EAST IN THE MEDIA

The Middle East in the Media
Conflicts, Censorship and Public Opinion

Edited by
Arnim Heinemann
Olfa Lamloum
Anne Françoise Weber

SAQI

in association with
Institut Français du Proche-Orient
Friedrich-Ebert-Stiftung
Orient-Institut Beirut

ISBN: 978-0-86356-658-5

© Arnim Heinemann, Olfa Lamloum and Anne Françoise Weber, 2009
Copyright for individual texts rests with the authors and translators

All rights reserved. No part of this book may be reproduced or transmitted in any form or by any means, electronic or mechanical, including photocopying, recording or by any information storage and retrieval system, without permission in writing from the publisher.

This book is sold subject to the condition that it shall not, by way of trade or otherwise, be lent, re-sold, hired out, or otherwise circulated without the publisher's prior consent in any form of binding or cover other than that in which it is published and without a similar condition including this condition being imposed on the subsequent purchaser.

A full CIP record for this book is available from the British Library.
A full CIP record for this book is available from the Library of Congress.

Printed and bound by Thomson Press (India) Ltd.

SAQI
26 Westbourne Grove, London W2 5RH
2398 Doswell Avenue, Saint Paul, Minnesota, 55108
Tabet Building, Mneimneh Street, Hamra, Beirut
www.saqibooks.com

in association with

www.ifporient.org

and

www.fes.de

and

www.oidmg.org

Contents

Acknowledgements 7
Introduction 9

Part One: Media in Conflicts

The Development of the Cartoons Crisis – A Danish Perspective 17
Jørgen S. Nielsen

The Representation of Middle East Conflicts in French Media 35
Denis Sieffert

Grapes of Unity facing Grapes of Wrath – War Coverage and Objectivity 45
Zahera Harb

Media War or War on Media: Lessons from Iraq 71
Ehab Bessaiso

Al-Jazeera: a Pan-Arab Revival? 98
Olfa Lamloum

Journalist as Change Agent – Government repression, corporate feudalism and the evolving mission of Arab journalism 116
Lawrence Pintak

Part Two: Market and Censorship

Is the Arab TV Viewer a King or a Pawn? How Arab Broadcasters Deal with Schedules and Audience Data 131
Naomi Sakr

The Business of News: One Writer's Impressions of
Two Middle East News Publications in 2005   147
Jim Quilty

Transnational Media and Authoritarian National Public Spheres   155
Tristan Mattelart

Representation, Images and Censorship in Algeria   172
Ghania Mouffok

Beyond Utopias and Dystopias: Internet in the Arab world   184
Maha Taki

A Reading of the Media Performance of the Muslim Brotherhood
in the 2005 Parliamentary Elections: A Case Study of the City of
Alexandria   195
Husam Tammam

Part Three: Public Opinion

Measuring and Comparing Opinions: A Practical and
Theoretical Challenge   211
Erik Neveu

European-Middle Eastern Relations in the Media Age   229
Kai Hafez

The Integration of Weblogs in the Egyptian Media Environment   252
Enrique Klaus

Do the Different Formats of the Lebanese Media constitute a
Pan-Arab Public Opinion?   268
Katharina Nötzold

In the 'Net' of Public Opinion – Islamist Online Media at Work   285
Yassin Musharbash

Contributors   297
Index   301

Acknowledgements

The editors intend this book to make a substantial contribution to an analysis of the media and its impact on the typical picture of 'the other culture', as well as to the widespread discussion on the multifaceted transnational interactions that illustrate and influence contemporary history and political relationships between the Western and the Arab world.

This book is the result of collaboration between three cooperating European institutions established in Beirut: the *Orient-Institut Beirut*, the *Institut Français du Proche-Orient* and *Friedrich-Ebert-Stiftung*. We would like to thank all the authors who have contributed to this book, as well as Samir Farah and Franck Mermier for their constant encouragement, and Barbara Drieskens, Karim Eid-Sabbagh and Viviane Kosremelli Salloum, who have assisted us with the French and Arabic translations. Furthermore, we would like to thank our colleagues Badia Baydoun, Britta van Erckelens, Wiebke Fleig and Ariela Gross, as well as our students Jessica Bodmann, Karoline Eickhoff, Jennifer Jasberg, Marcel Lippert, Anna Seeger and Miriam Younes, who committed themselves during their internships at the *Orient-Institut* to this project and its publication.

The realisation of the project has been possible thanks to the encouragement, shared responsibilities, joint commitment and friendly cooperation of the three organizing institutions. The project has been supported as well by the *Fritz-Thyssen-Stiftung*, the *Agence universitaire de la Francophonie*, the *Ecole Supérieure des Affaires* in Beirut and our patient and understanding partners, Lucy Chapman, Dina Dally, Lara Frankena, André Gaspard and Shikha Sethi at Saqi Books.

Editorial Note

Arabic words have been transcribed in a simplified way, as have most names, except those well known by the public through the press.

The Editors

Introduction

Several long-lasting conflicts shape the Middle East and dominate its representation in local, regional and international media. By September 2001, the implications of framing and agenda-setting reality in the media, including the Internet, have even more fatefully than before linked pictures of 'the other culture' in the West and the Middle East. The coverage on either side is biased in part because of direct censorship in media systems, which is often due to the impact of political and religious authorities. Especially in times of increasing use of the Internet, the role of online censorship in controlling a channel for mostly oppressed voices is, within this context, an important issue, although there are discourses avoiding control and restriction.

In addition, the political stand of the reporter and his employer condition the way in which the conflicting parties and the stakes of conflict are presented. Specific forms of expression and images are in use to influence the recipient's opinion more or less unconsciously. In the West, as well as in the Arab world, economic conditions structuring media production and performance have an enormous impact on regulating and conditioning the ways conflict is presented.

Over the last century, media has become an even more powerful tool, its link to public opinion strengthening as the breadth of media grows. In this context, media is a tool of analysis, as it reflects public opinion by viewing the world through the prism of its consumers, providing images and broadcasts that its consumers want to see, hear and read. In this respect, media is also a tool of persuasion, as it shapes public opinion by disseminating the government's, a religious or political party's, or a stakeholder's position. The discussion of this functional relationship shows that media also challenges the public opinion to be active in its own creation, as is the case with the emergence of the Internet. Yet, different media channels address different publics and reflect different agendas. Proof is given through the varied emphasis on themes covered by the large media houses in the West and in the Arab world.

The focus of the discussion reflected in this publication tends towards an analysis of the Arab media sphere, as well as its reflection and response in Western media.

We are far from the time when the full scope of Arab media was easy to grasp; when it was limited to local TV, public radio and a press under the control of the elite in power, and when rare discordant voices were obliged to take shelter in Beirut, London or Paris. In fact, since the mid 1990s, the Arab media space has become progressively complex, thus mobilising a variety of individual or collective actors, dissidents or opponents to the government. The appearance of over 400 Arab TV satellite networks, owned by sixty companies;[1] the outbreak of Salafist satellite channels; the launch at the beginning of 2008 of the BBC non-stop information channel in Arabic, consequent to German, Russian and American channels; the multiplication since the fall of Saddam Hussein's regime of Iraqi television stations held by Shiite Islamist organisations,[2] as well as those promoted by *hawza* (Shiite religious schools) or *marja'iyya* (Shiite religious authorities); the visibility of young bloggers in Egypt and their marked implication in the general strike of April–May 2008;[3] the occurrence of digital radios in Jordan and in the occupied Palestinian territories; not to mention jihadist websites, are but some of the recent examples of the complexity of the rich and developing media universe. The stakes are such that in spite of the divisions shaking Arab regimes, the latter succeeded in February 2008, upon Egyptian and Saudi Arabian initiative, in achieving an almost unanimous solution to one issue, with the adoption of an agreement regulating satellite, radio and TV broadcasts in the Arab world.[4] Consequently, three Egyptian channels – al-Baraka, al-Hikma and al-Hiwar – found themselves sanctioned, and were forbidden from broadcasting via NileSat.[5]

This new configuration of the media space has opened unique research paths probing new modes of production, broadcast, suitability and censorship of information in the Arab space. The power of these questions

Introduction

was reinforced after the war against Afghanistan. This is nothing to be surprised about bearing in mind that this war, led by the United States, saw the emergence, for the first time, of an Arab supplier of images and sounds, altering the United States's monopoly on information.

More globally, the Arab media erupted, breaking through the global media field under circumstances of conflict and war. Al-Jazeera, a 24-hour information channel launched in 1996, found international authentication in 2001 during its controversial coverage of the first 'war against terrorism' in Afghanistan. As for al-Manar, the channel of the Lebanese Hizbullah, it became a political entity, to the extent that it is classified as a terrorist organisation by the Bush administration, and banned in Europe because of attitudes towards the Israeli-Palestinian conflict judged to be anti-Semitic.

Moreover, the interest stimulated by the Arab media is relatively recent. In social sciences, research carried out on this topic was starting to see the light by the end of the 1990s and was still scarce compared with that undertaken on Islamist movements, for example.

Al-Jazeera's breakthrough in the Arab media field focused the interest of initial research into this issue. Subsequently this phase, both necessary and legitimate, was gradually replaced with new investigation. Al-Jazeera had sufficient merit to offer us a stimulating observation post from which to broaden our perspectives. Furthermore, it contributed greatly to a new aspect of research, still developing, of 'media in the Arab space'. However, it is definitely the war in Iraq of 2003 that opened new horizons to research. The numerous faces in the war against Iraq, showed in the media field – including that of the Arab world – and the resumption of conflict in the region, generated 'media wars' and renewed investigation into the shape and mechanism of the representation of conflict and its incidence on social relations.

The premise of this publication is confluent with the fruits of three international colloquiums organised in Beirut between 2005 and 2007: the first Beirut Media Forum was entitled *Media and European-Middle Eastern Relations. A Virtual Dialogue?*; the second was *Media and the Public Opinion between Europe and the Middle East*; and the last one was *Middle Eastern Conflicts in the Media – Censorship and Representation*. This book assembles a selection of contributions promoting a multi-disciplinary approach, comparing academic analysis to practical experiences of media professionals.

This concept, which also structures the book, is the deliberate reasoning behind the diversity of its articles, and is regarded as a double balance: a balance of Western and Arab perspectives, as well as the balance of an approach academic on one hand and journalistic on the other. Readers interested in the relationship between the media and culture, particularly in an era that is defined by information, communication and globalisation, will gain many useful insights in this work. The book addresses a broad public concerning itself with patterns of historical and cultural development in the Arab and Western media context.

The book is structured in three parts. The first part tackles the issue of media representation and restitution through words and images about conflict. It deals with the foundations and consequences of journalistic writing on crisis perception, its intelligibility, and even its evolution in a comparative approach, looking at Europe, the United States and the Arab world.

The second part deals with different forms of censorship and restriction – political, through markets or through unequal access to technology – which alter the production, as well as the circulation, of information. It discusses the ability of transnational communications to get around the internal policies of censorship. Finally, it presents a case study on the achievements of the Muslim Brotherhood in bypassing censorship during the last legislative elections held in Egypt.

The last part reconsiders the issue of the public sphere in the Arab space, through the exploration of theoretical and practical challenges suggested by the appreciation of 'public opinion'. It tackles the impact of new trans-national media on relations between Europe and the Middle East. Finally, it discusses three case studies – weblogs in the Egyptian media context, amusement programmes on Lebanese networks, and e-jihadism – addressing the theory of training, new mobilisation or identification spaces.

Introduction

Notes

1. This number was disclosed during a meeting of Arab information ministers in February 2008.
2. A German NGO, Media in Cooperation and Transition (MICT), made an inventory of at least eight official channels linked to Islamist associations, among which were Badr Organisation (2), the Islamist party Al-Da'wa (2), and the Supreme Islamic Council (2). Cf. *Media on the Move: A reader on Iraqi Media and Media Law*. MICT, 2007.
3. Cf. Al-Masri Al-Yawm, Egypt, April 18, 2008 on Facebook mobilisation.
4. Lebanon opposed the agreement, whereas Qatar abstained. However, according to some observers, one of the most repressive provisions of the document concerned the criminalisation of the viewers.
5. Cf. Al-Quds al-Arabi, 22 April 2008.

PART ONE

Media in Conflicts

JØRGEN S. NIELSEN

The Development of the Cartoons Crisis – A Danish Perspective[1]

A chronology of the affair

The publication of twelve cartoons depicting the Prophet Muhammad in a Danish national newspaper was hardly noticed at the time – and no one could have predicted how the ripples would ultimately turn into shock waves.

The initial publication had been decided explicitly to test the bounds of press freedom, after a Danish author writing a children's book about Muhammad announced that he had been having trouble finding someone willing to illustrate it.[2] In the 30 September 2005 edition of *Jyllands-Posten*, the culture editor of the paper, Flemming Rose, referring to recent instances of artistic self-censorship in relation to Islam, wrote:

> Modern secular society is rejected by some Muslims. They are demanding a special position when they insist that special consideration should be given to their religious sensitivities. This is incompatible with secular democracy and freedom of expression, in which one must be prepared to tolerate derision, mockery and ridicule.[3]

This was accompanied by twelve cartoons depicting the Prophet Muhammad drawn by twelve different artists, including the most famous one in which the prophet is depicted with a bomb in his turban.

The cartoons aroused anger among many Danish Muslims, some of whose organisations appealed to Arab and Muslim embassies when the Danish authorities failed to respond to their complaints. The lead in the protests was quickly taken by an association of Muslim organisations called *Det Islamiske Trossamfund*, the 'Islamic Faith Community'.

On 12 October eleven ambassadors of Muslim countries sent a joint letter to Prime Minister Anders Fogh Rasmussen:

> This pertains to an ongoing smear campaign in Danish public circles and media against Islam and Muslims. Radio Holger's remarks, for which it was indicted, DF (Dansk Folkeparti) MP and mayoral candidate Louise Frevert's derogatory remarks, Culture Minister Brian Mikkelsen's statement on war against Muslims and *Daily Jyllands-Posten*'s culture page inviting people to draw sketches of Holy Prophet Muhammad (PBUH) are some recent examples.
>
> We strongly feel that casting aspersions on Islam as a religion and publishing demeaning caricatures of Holy Prophet Muhammad (PBUH) goes against the spirit of Danish values of tolerance and civil society.
>
> In your speech at the opening of Danish Parliament, Your Excellency rightly underlined that terrorists should not be allowed to abuse Islam for their crimes. By the same token, Danish press and public representatives should not be allowed to abuse Islam in the name of democracy, freedom of expression and human rights, the values that we all share.
>
> We deplore these statements and publications and urge Your Excellency's government to take all those responsible to task under the law of the land and in the interest of inter-faith harmony, better integration and Denmark's overall relations with the Muslim world. We rest assured that you will take all steps necessary.
>
> Given the sensitive nature of this matter, we request an urgent meeting at your convenience. An early response would be greatly appreciated.[4]

On 19 October the prime minister refused to receive the eleven Muslim ambassadors. He chose to focus solely on the reference to the cartoons stating that, as this was a question of freedom of the press, there was nothing to discuss.⁵

The Egyptian government in particular took offence at the Danish prime minister's refusal to meet with the ambassadors. It raised questions about the Danish government's seriousness in wishing to hold a dialogue with the Arab world through its Arab Initiative, started in 2003. A significant aspect of this programme was the establishment of the Danish-Egyptian Dialogue Institute in Cairo in January 2005 as a joint project between the two foreign ministries. The Danish Muslim organisation *Islamisk Trossamfund* sent a delegation to various Arab countries to raise a protest, carrying with them a file containing not only the original cartoons but also other unrelated offensive material, as well as general information about the position of Islam in Denmark.⁶ At this stage the protests were looking forward to the Islamic summit due in early December.

The summit conference of the Organisation of the Islamic Conference (OIC), meeting in Mecca on 5–6 December 2005, discussed the issue of Islamophobic material being published in the western media including Denmark and included reference to the cartoons case in the final communiqué, although without mentioning Denmark:

> The conference expressed its concern at rising hatred of Islam and Muslims and condemned the recent incident of desecration of the image of the Holy Prophet Mohammad (PBUH) in the media of certain countries and stressed the responsibility of all governments to ensure full respect of all religions and religious symbols and the inapplicability of using the freedom of expression as a pretext to defame religion.⁷

By late December, the OIC and other international Islamic organisations began to break off relations with Danish organisations they had previously been cooperating with. In particular this affected the plans of the Danish Centre for Culture and Development (DCCD), plans financially supported by the government, for a major cultural festival called 'Images of the Middle East' due to take place June–September 2006 in a number of venues across the country. The OIC withdrew from a cooperation agreement with the DCCD, and the Islamic Educational Scientific and Cultural

Organisation (ISESCO) asked that joint projects be postponed until the situation calmed down.

The Danish prime minister's New Year speech and a telephone conversation between the foreign minister and 'Amr Musa, secretary-general of the Arab League, seemed to calm down the situation. In his New Year speech, Anders Fogh Rasmussen condemned

> any expression, action or indication that attempts to demonise groups of people on the basis of their religion or ethnic background. This sort of thing does not belong in a society that is based on respect for the individual human being.[8]

At the same time he repeated his now-standard defence of the freedom of expression. The latter part of the speech was widely circulated in both English and Arabic translations by the Danish foreign ministry and by Danish embassies in the Muslim world. The prime minister's phrase, quoted above, was repeated when the foreign minister, Per Stig Møller, wrote to 'Amr Musa on 6 January 2006 following an extended telephone conversation earlier that day. He also repeated the defence of free speech with the phrase 'Freedom of expression is absolute'.[9]

On 10 January a small conservative Christian magazine in Norway, *Magazinet*, republished the cartoons, sparking renewed crisis as newspapers and magazines in other countries also published them. On 21 January the International Union of Muslim Scholars, whose president was Dr Yussuf al-Qaradawi and general secretary Dr Muhammad Selim al-Awa, called for a boycott of Danish goods,[10] a call first taken up a few days later in Saudi Arabia which then spread quickly across the Arab and many other parts of the Muslim world.[11]

The situation was now critical for both Danish trade and for the country's political standing. On the evening of 30 January 2006, Claus Juste, the editor of *Jyllands-Posten*, published a statement in Danish, English and Arabic seeking to explain what the intention of the cartoons had been and apologising for the unintended hurt caused. He rejected the accusation that the cartoons were part of a campaign against Islam and Muslims, as had been alleged, and asserted his paper's continuing support for freedom of expression, communal harmony and mutual respect.[12]

The Danish prime minister immediately issued his own statement

The Development of the Cartoons Crisis – A Danish Perspective

referring to the apology, stressing that he 'deeply respected the religious feelings of other people', and expressing his distress 'that these drawings have been seen by many Muslims as a defamation of the Prophet Mohammed and Islam as a religion.'[13] He followed this up with an interview broadcast on the Arabic satellite-TV station Al Arabiya on 2 February.[14] Optimistically the TV station publicised the interview in advance as containing the hoped-for apology which, in the event, it did not.

By this time the situation was out of control. Demonstrations took place in many Muslim countries, some violent, leading to a number of deaths and to attacks on various embassies including attacks on the Danish embassies in Damascus, Beirut, Tehran, Jakarta and Islamabad. For a few weeks the crisis was in deadlock, with one side demanding an apology and the other side offering explanations and regrets but no apology. The crisis was also becoming internationalised with talk of a complaint to the UN Commission on Human Rights, especially when the commission's special *rapporteur* on contemporary forms of racism, racial discrimination, xenophobia and related intolerance, Doudou Diène, accused the Danish government of failing to show the commitment and vigilance that it normally displays in combating religious intolerance and incitement to religious hatred and promoting religious harmony.[15]

On the other hand there was a growing realisation that some form of dialogue had to be started. On 22 February, the Danish foreign minister announced a 'series of forward-looking initiatives aimed at promoting respectful dialogue'.[16] With support from the ministry a hectic dialogue programme started. Delegations from Denmark, mainly consisting of government officials, ministers and bishops, met with delegations from various Arab countries, in turn consisting of civil servants, ministers and muftis. Initially these meetings took place in neutral locations like Geneva or Vienna – at the time Austria held the rotating presidency of the EU Council of Ministers – but soon also in Copenhagen and in Arab capitals.

The peak of the troubles passed during late February, and by the end of March other things were happening in the Arab world to distract attention. The 15 March decision by the Danish public prosecutor not to take *Jyllands-Posten* to court for blasphemy[17] did not spark the expected further protests. The boycott gradually weakened until it was officially called off in Saudi Arabia in May. Although the demanded apologies have not been forthcoming, energies appear to have been dissipated in the streets, and by the end of April the issue was merely simmering. However, Danish exports suffered significant losses as a result, falling by as much as 88 per cent in Libya between March and June 2006, 47 per cent in Iran and 40 per cent in Saudi Arabia. By the anniversary of the publication of the cartoons, estimates of total export losses were in the region of one billion Danish kroner.[18]

A Danish 'Rushdie affair'?

It has been suggested that the affair bears many similarities to the Rushdie affair of 1990, when Muslims campaigned against the publication of Salman Rushdie's *The Satanic Verses*. To a certain extent comparisons can be made.

The demographics of the Danish Muslim community today are similar to that of the UK Muslim community then, in that the first major tranche of Muslims born in Denmark has come into the labour market in recent years and experienced discrimination – the family reunion process among immigrants in Denmark having started twelve to fifteen years later than in the UK.

The Danish political elites have taken little notice of the growing Muslim community in Denmark, and public awareness has been based on hearsay and media stereotyping rather than direct experience. This has allowed some politicians, and especially those of the right-wing Danish People's Party, to set an anti-Muslim tone to which Muslims have found it difficult to respond. Certain Muslim groups have sought to make political capital, domestically and internationally, out of the crisis.

The crisis ran out of control when it became internationalised and exploited for various political purposes. In 1990 the context was Iranian domestic and regional politics. This time it was tensions between authoritarian governments and Islamic political movements using democratic processes to gain influence, in particular after the electoral successes of the Muslim Brotherhood in Egypt and Hamas in Palestine.

In response to the strengthening of the political extremes both in wider society and among Muslims, there is a counter-process of mobilisation of people in the mainstream in both sectors, as witnessed by a number of demonstrations and citizens' initiatives involving Muslim and broader Danish groups, secular as well as Christian.

The Development of the Cartoons Crisis – A Danish Perspective

The debate has again been about the tensions between values claimed by a democratic and liberal Europe and those claimed by communities with strong religious identities, focused on the contest between freedom of expression and respect for religious symbols.

There are, of course, limits to the comparison, first and foremost due to the developments which have taken place in the decade and a half between the two affairs, domestically and internationally.

Domestically, within Europe, the process of settlement and integration has moved on, with both positive and negative aspects. On the one hand have been integrative moves in various countries, including growing political participation on the part of Muslims in most countries, France being a significant exception. Germany has eased access to citizenship for its Turkish residents, and in many countries Muslims have been involved in provision of social, educational and other services. Some countries have tightened up legislation against incitement to racial hatred and against hate speech, and in some cases attempted to expand coverage to incitement to religious hatred. There have been a number of initiatives in the media to spread more and better information about Islam and Muslims, and religious education syllabuses have become more inclusive in many countries.

On the other hand, there has been a spread of so-called Islamophobia, a contested term denoting expressions and actions which often are basically racist but find cover behind a focus on Islam, at a time when there is a public perception that Muslims are fair game. In the current case, it has been suggested that often the argument for freedom of the press is a cover for xenophobia, an accusation that has been levelled at the Danish newspaper which published the original cartoons, particularly graphically in a cartoon published in the British Sunday newspaper *The Observer* on 5 February, depicting a big sinister figure marked 'xenophobia' hiding behind a frightened torch-bearing woman marked 'freedom of speech'.

Playing very strongly into the domestic scene has been the so-called 'war on terror', mainly targeted at terrorism originating in Muslim networks. This theme developed during the 1990s and was sharply focussed by the attacks of 11 September 2001 and reinforced since by the bombings in Madrid and London. Muslims have come to feel that government policies and public attitudes to them are now security-led. Internationally, many Muslims – and not just the religiously active ones – are convinced that the 'war on terror' is actually a war against Islam. It is no coincidence that the cartoon which caused the most offence was the one showing Muhammad wearing a bomb in his turban.

These elements play out differently in different European countries, depending on local circumstances. The violence seen in demonstrations in the Muslim world has not been repeated in demonstrations in Europe. A small demonstration in London exhibiting violent slogans provoked a much larger, peaceful Muslim demonstration expressing its disgust at those slogans. In France, the violence of the autumn riots was not repeated, and Muslim organisations have attempted to use the courts. Policies of integration are still contested, but they apparently continue to show slow progress.

The cartoons events have again raised questions of what constitutes citizenship, first raised at the time of the Rushdie and headscarf affairs in 1990 in Britain and France respectively. The actions of the majority of the Muslim inhabitants in most countries would seem to indicate that they regard themselves as citizens but want to negotiate some of the details, above all the place of religion in public life. The actions of some governments, including the Danish, seemed to indicate that they have not been ready to enter into that negotiation.

Religion in the public space

In an interview with Denmark Radio on 15 February 2006, Prime Minister Anders Fogh Rasmussen argued for religion to stay out of the public space.

> We have to be careful that religion does not take up too much of the public space if Danish society is to continue to cohere. Lately, I think religion has come to take up too much space in the public debate, and I would emphasise that I respect every individual's religious conviction. But I really think that it is a quality which holds Danish society together that we have a tradition of not mixing, or at least differentiating between, politics and religion. And that religion is assumed to be a personal matter.[19]

The problem with his argument was that one of the main themes throughout the crisis had been considerations of the pressure from the right-wing Danish People's Party (DPP). The party was founded during the 1990s around a programme of defending the national character and culture of Denmark against foreign, and especially non-European, influences. Central to this character was Denmark's Lutheran Christian tradition:

> Christianity has been upheld for centuries in Denmark and is inseparable from the life of the people. The significance which Christianity has had and continues to have is enormous and leaves its stamp on the Danish way of life. Over the ages it has been the guide and set the norms for the people.[20]

In fact, one of the key movers in the establishment of the party had been a pastor in the Danish state church, Søren Krarup, whose brother-in-law Jesper Langballe, also a pastor, joined him as a member of parliament for the party.

The current Danish government is a minority coalition government led by the prime minister's party Venstre (or 'Liberal') with the Conservatives as the junior partner. The coalition primarily depends on the DPP to support its legislative programme through parliament. It came to power in a general election held in October 2001 after an election campaign in which immigration and integration were the most contested issues, a campaign in which the attacks of 11 September on New York and Washington served to focus this issue on Muslims.[21] Subsequently, the new government introduced a sharp tightening of immigration law, to the extent that there were accusations from Strasbourg that Denmark might be in breach of the European Convention on Human Rights. Some analysts would judge that the prime minister is on the right flank of his party and that he sympathises with many of the DPP's attitudes. Certainly the tightening of legislation against immigrants has had the full support of the DPP, and on a number of occasions the prime minister has rapped the knuckles of ministers who have incurred the displeasure of the DPP.

The anti-immigrant and anti-Muslim character of much public debate contributed to a sense that various forms of anti-Islamic expression had become legitimate, although such comments on the private Holger radio station led to a successful prosecution, and particularly harsh comments by a DPP member of parliament on her personal website had to be withdrawn.

The debate had also been such that many Muslims felt it wise to keep a low profile. This had the effect of leaving the field open for more robust *Salafi-*oriented organisations, particularly *Islamisk Trossamfund*, to monopolise the public representation of Islam in ways that often served to confirm the image of Islam as incompatible with Danish values which has been promulgated by the DPP.

It is this background to which the ambassadors referred in their letter to the prime minister of 12 October 2005. It seems reasonable to judge that this was a discussion that the prime minister did not wish to enter, for which reason it was decided to draw attention away from the substance of the letter to the sole issue of freedom of speech.[22]

Behind all the trouble lies an uncertainty over Danish self-identity of a magnitude that deserves the epithet *Kulturkampf*. Is Denmark going to participate in the wider world, other than through trade, or is it going to restrict itself to sitting on the sidelines defending its past? This is possibly a bit harsh. After all, Denmark has had a commendable internationalist history since 1945 as an active member of the United Nations and as one of a very few countries long since to have met the UN's agreed level of contribution to development aid. But that was the world out there. Since the 1970s, that world has increasingly entered into Denmark, and much of the country has defended itself by turning to the imagined resources of 'Danishness'. Denmark's Euro-scepticism rests on very different foundations than does that of Britain. An increasingly important part of those foundations is the country's Lutheran Christian identity. But this is also a contested area. Many remain secular and desire to keep religion out of the public space altogether. Others want to make space for religion in public, but they are split between those – Christians and Muslims – whose claims are absolute, who dominated the discourse during the crisis, and those whose approach is inclusive. Again, the DPP's Statement of Principles highlights the point of contest:

> Denmark is not a country of immigration and never has been. We will therefore not accept that the country become multi-ethnic.[23]

In early February, Uffe Elleman-Jensen, a former leader of the party now led by the current prime minister, took the opposite stand very firmly:

If we Danes wish to preserve dialogue with other cultures and religions – and even wish that they buy our milk products – then we cannot demand that they all accept our norms, least of all when they are exposed to disdain, mockery and sarcasm. If we insist that they have to tolerate that, we are all firmly anchored in 'the Danish village pond' where everybody is convinced of her/his own infallibility and therefore not able to get on in a globalised world.[24]

Legal issues, the media and tunnel vision

Several themes have run through the events of the cartoons crisis – and perhaps someone can also make something out of the fact that the usual term in Danish is the 'Muhammad crisis'. There are at least questions of perception of the legal situation, of the way the media responded and of the selective perspectives of the various parties.[25]

From the outset, two different aspects of Danish legal concepts were contested in the debate. On the one hand, mobilised in particular by the prime minister, *Jyllands-Posten* and its sympathisers and the Danish People's Party, was the principle of freedom of expression. On the other hand was the question of blasphemy, argued by many of the Muslim participants.

The argument mounted at various times by government ministers and other politicians, as well as by defenders of *Jyllands-Posten*, that freedom of expression is a core value of Danish democracy, was often phrased in terms of it being absolute. A common expression in the public debate was that 'freedom of expression is non-negotiable'. Prime Minister Anders Fogh Rasmussen set the tone early on and was quoted in *Jyllands-Posten* on 30 October 2005 as saying, 'It is very difficult to draw a line by legislation, and it is a loss for democracy if we place any kind of limitation on the freedom of expression.'[26] Here was clearly an instance of a constitutional principle being exaggerated for political purposes. Paragraph 77 of the Danish constitution states:

> Any person shall be at liberty to publish his ideas in print, in writing and in speech, subject to his being held responsible in a court of law. Censorship and other preventive measures shall never again be introduced.[27]

Accountability to the courts includes offences such as libel and slander, incitement to violence, fraudulent claims, etc. Similarly, employees are limited in what they can say about their employer's activities by contractual terms of confidentiality, a fact which was ironically demonstrated when the Danish ambassador to Saudi Arabia spoke more freely about the cartoons than the foreign ministry had authorised, as a result of which he was reprimanded.

Limitations to freedom of expression are also explicitly envisaged in the European Convention on Human Rights, which Denmark was among the earliest to ratify.[28] Many participants in the public controversy referred to Article 10 of the Convention:

> Everyone has the right to freedom of expression. This right shall include freedom to hold opinions and to receive and impart information and ideas without interference by public authority and regardless of frontiers. This Article shall not prevent States from requiring the licensing of broadcasting, television and cinema enterprises.

But, as Larsen and Seidenfaden point out,[29] they tended to ignore section 2 of the same article:

> The exercise of these freedoms, since it carries with it duties and responsibilities, may be subject to such formalities, conditions, restrictions or penalties as are prescribed by law and are necessary in a democratic society, in the interests of national security, territorial integrity or public safety, for the prevention of disorder or crime, for the protection of health or morals, for the protection of the reputation or rights of others, for preventing the disclosure of information received in confidence, or for maintaining the authority and impartiality of the judiciary.

This takes us directly into the other half of the argument, namely the question of blasphemy. According to section 140 of the Danish criminal code,

> Public ridicule or mockery of the religious beliefs or worship of a legally existing faith community is punishable by fines or imprisonment of up to four months.[30]

According to the law, therefore, the section represents a limitation of freedom of expression. This limitation has been explicitly supported by the European Court of Human Rights as recently as 1996 in the case of *Wingrove v. United Kingdom*.[31] Following the murder in the Netherlands in 2004 of the Dutch film director Theo van Gogh, some Danish politicians had suggested that the blasphemy paragraph should be abolished. Proposing this to parliament in March 2005, the Danish People's Party's theological argument was:

> From the point of view of principle and of religion, a law which punishes blasphemy is a total misunderstanding in a Christian country. The idea that we should protect God's honour with law in hand is almost blasphemous.[32]

But as Henning Koch, professor of law at Copenhagen University, shows, this represents a serious misunderstanding of the nature of blasphemy in Danish law.[33] He argues that since the end of the nineteenth century the legislation against blasphemy has been primarily concerned with the preservation of public order and has been located in that chapter of the criminal code since 1930. The law on blasphemy was last successfully used in Denmark when a group of Nazis were found guilty for their attacks on Jews by way of serious misrepresentation of Jewish teachings.

In tandem with the section on blasphemy is section 266b, the so-called 'racism' section, which criminalises expression that threatens, mocks or derides a group of persons 'because of its race, skin colour, national or ethnic origin, faith ...'[34] Professor Koch argues that, if the blasphemy section were abolished, this section could be circumvented, as the Nazis did in 1938, by shifting the target of attack from a group of people to core aspects of their belief or practice. But, he stresses, referring to the European Convention and its case law, the importance of the basic right of free speech is such that only very serious breaches of sections 140 and 266b should be regarded as punishable by the courts.

On 15 March 2006, the director of public prosecutions published his decision that *Jyllands-Posten* did not have a case to answer in relation to the two sections of the criminal code.[35] He thus confirmed a similar decision published at the beginning of January by the public prosecutor for the eastern region of Denmark. The matter had been appealed to the director with the complaint that, in this instance, reference to the principle of freedom of expression as laid down in the constitution was not justified. The decision effectively closed the recourse to law, even if it did not close off the legal argument. It is certainly arguable that the director of public prosecutions might have been wiser in this instance to let the courts decide, given the public significance of the case, rather than assuming quasi-judicial authority himself.

The fact remains that the polemics both in the media and among the politicians ran roughshod over the law and, many would suggest, basic facts, not to mention engaging in grossly partisan and self-interested misrepresentation. It started with the government's decision to respond to the ambassadors' letter of 12 October 2005 by focusing solely on the issue of 'absolute' freedom of expression. As minister also for the press, the prime minister was the most obvious member of the government to at least discuss the issue even if he had no authority to exercise any form of censorship. The response ignored the main point made in the letter concerning the general development of anti-Islamic discourse in Danish debate over a number of years, but especially since the autumn of 2001. But the letter itself chose to play down the responses to at least two of the cases mentioned. Radio Holger was not only indicted, it was found guilty under section 266b of the criminal code and the owner punished. The material published on the DPP MP's website had provoked such an outrage that she had been forced to apologise and to take the material off. There was clearly something to discuss here between both parties to the correspondence.

The careless treatment of facts continued in the presentations of the case circulated by some of the Muslim organisations, in which their situation in Denmark was painted rather more negatively than the facts warranted.[36] In particular, it was noted that Muslims had, despite years of campaigning, still not got a central mosque in Copenhagen. The file taken to the Middle East by a delegation of Muslims failed to mention that a plot of land had been set aside for precisely this purpose by Copenhagen's city council but that Muslims had been unable to agree among themselves on the ownership, management and financing of the project. It also stated correctly that Islam had not become a recognised faith community but failed to note that it had been 'approved'. It was for most practical purposes on an equal footing with the recognised religious communities.[37]

In the arguments there were numerous instances of selective memory. One of the most obvious was the fact that *Jyllands-Posten* in April 2003 had turned down a cartoonist's offer of drawings making fun of the resurrection of Jesus. The offer was refused with the words, 'I don't think the readers of *Jyllands-Posten* will find them amusing. In fact, I think there would be an outcry.'[38]

Among most parties in the case, there was clearly a failure of imagination. Hardly anyone in Danish media, political or government circles foresaw the possible reactions among Muslims at home or abroad. Very few in government had imagined the consequences of the decisions taken several years earlier to adopt a much more proactive foreign policy outside Europe, and especially in the Middle East, alongside the United States and the United Kingdom. Similarly, the Muslim delegation taking their case to the Arab world probably did not imagine the consequences of their actions either, although they had been warned.[39]

Today in Denmark the case rumbles stubbornly on as the public debate has become almost obsessed with religion – Islam, Christianity, any religion, and questions of religion and politics, religion in the public space. Ulla Holm's comment in February 2006 – 'The impact of the publication of the cartoons may now reconstruct what Danish identity is about'[40] – looks as if it is being confirmed.

Postscript

Since the cartoons crisis, there has clearly been a significant evaluation of the experience. While some on the right are stubbornly sticking to their views, others have moved on.

At the time of the 2006 war between Israel and Hizbullah, the Danish government acted with commendable speed and effectiveness to evacuate over 5,000 Danish citizens, virtually all of Arab origin, from Lebanon. The chair of the Danish Palestinian Friendship Society noted that, from all quarters, the term used about the evacuees was not the usual 'new Danes' or some form of ethnic distinction like Arab, Lebanese or Palestinian Danes, but simply 'Danes'.[41] There were the inevitable voices of dissent from parts of the Danish People's Party, whose MP Søren Espersen suggested that the returning evacuees should be checked by the social security offices to see if they were receiving welfare payments during their absence in Lebanon, a suggestion which was met with almost universal condemnation.

The evacuation was in part an answer to accusations of incompetence in responding to the plight of Danish holidaymakers in Thailand at the time of the Indian Ocean tsunami at the end of 2004. But it was also a response to accusations during the cartoons crisis that the country treated Danes of Arab origin as second-class citizens.

In October 2006 a second cartoons crisis threatened, when a video was broadcast showing the youth wing of the Danish People's Party competing to produce the most insulting drawings against Islam at a summer camp in early August. Very promptly the prime minister issued a strong condemnation, and the youth wings of all the other parties cut off cooperation with that of the DPP, while the DPP itself disowned the action. At the same time, Muslim organisations moved fast to calm the waters as the story started circulating in the Muslim world.

Notes

1. This chapter was presented at the conference on 'Public opinion and the media between Europe and the Middle East', organised by the German Orient Institute, FES and IFPO, Beirut, 15-17 November 2006. An earlier version of part of this chapter was presented at the annual conference of the British Society of Middle East Studies, Birmingham, 24 July 2006.
2. The book was published in spring 2006 as *Koranen og profeten Muhammeds liv*, Copenhagen: Høst.
3. Flemming Rose, 'Muhammeds ansigt', *Jyllands-Posten*, 30 September 2005; translations from Danish are the author's unless otherwise stated.
4. Reproduced in *Politiken*, 21 February 2006.
5. *Politiken*, reviewing the case on 12 and 19 February 2006, suggested that the foreign ministry's translation of the ambassadors' letter had inadvertently made it demand that the prime minister take *Jyllands-Posten* to court; see Klaus and Mikael Rothenstein, *Bomben i turbanen*, Copenhagen: Tiderne Skifter, 2006, p. 33.
6. The copy I am working from is a revised version which includes material from visits to Egypt which took place in November and early December 2005.
7. Final communiqué of the third extraordinary session of the Islamic summit conference, [http://www.oic-oci.org/ex-summit/english/fc-exsumm-en.htm (accessed 1 May 2006)].
8. Full text accessed 5 January 2006 on [http://www.stm.dk].
9. Personal copy from confidential source.
10. Accessed 27 January 2006 on [http://qaradawi.net].

11. See report on *Arab News* 27 January 2006, accessed 27 January 2006 on [http://www.arabnews.com].
12. Accessed 31 January 2006 on [http://www.jp.dk].
13. English copy supplied by the Danish embassy in Damascus.
14. Danish text accessed 3 February 2006 on [http://www.stm.dk].
15. UN Economic and Social Council, Commission on Human Rights, 62nd session, doc. E/CN.4/2006/17, 13 February 2006, p. 10.
16. English text supplied by the Danish embassy in Damascus.
17. Published by the office of Rigsadvokaten on 15 March 2006 as act no. RA-2006-41-0151, accompanied by a press release in English, accessed 15 March 2006 on [http://www.rigsadvokaten.dk].
18. *Politiken*, 30 September 2006.
19. Quoted in Klaus and Michael Rothstein, *Bomben i turbanen*, p. 155f.
20. From the DPP's Statement of Principles (Principprogram) accessed 9 July 2006 on [http://www.danskfolkeparti.dk].
21. See C. Allen and J. S. Nielsen, *Summary Report on Islamophobia in the EU after 11 September 2001*, Vienna: European Monitoring Centre on Racism and Xenophobia, 2002.
22. This point was made by one of the researchers taking part in the seminar on 'Moving beyond stereotypes' organised in Copenhagen 16–18 May 2006 by the Danish Centre for Culture and Development, ref. the report of the meeting published May 2006, pp. 8–9.
23. From the DPP's Statement of Principles (Principprogram) accessed 9 July 2006 on [http://www.danskfolkeparti.dk].
24. *Berlingske Tidende*, 8 February 2006, translated in Ulla Holm, 'The Danish ugly duckling and the Mohammed cartoons', *DIIS Brief*, February 2006.
25. Another set of issues revolves around the relationships between race, religion, multicultural society, and liberal democracy, issues which are debated by Tariq Modood, Randall Hansen, Erik Bleich, Brendan O'Leary and Joseph Carens, in 'The Danish cartoon affair: free speech, racism, Islamism and integration,' *International Migration*, 2006, no. 44 (5), pp. 3–62.
26. R. E. Larsen and T. Seidenfaden, *Karikaturkrisen: En undersøgelse af baggrund og ansvar*, Copenhagen: Gyldendal, 2006, pp. 236–8.
27. *The Constitutional Act of Denmark of June 5, 1953*, Copenhagen: Folketinget, 1999.
28. *European Convention on Human Rights: collected texts*, Strasbourg: Council of Europe Press, 1994, p. 62.
29. Larsen and Seidenfaden, p. 237f.
30. Cited in *ibid*. p. 237.
31. Ovey and White (eds), *Jacobs and White, The European Convention on Human Rights*, 3rd ed., p. 285f., and Gilles Dutertre, *Key case-law extracts: European Court of Human Rights*, Strasbourg: Council of Europe Publishing, p. 324f.

32. Cited in Henning Koch, 'Ytringsfrihed og tro', in Lisbet Christoffersen ed., *Gudebilleder: Ytringsfrihed og tro i en globaliseret verden*, Copenhagen: Tiderne Skifter, 2006, p. 84, note 4.
33. Henning Koch, pp. 72–88.
34. Ibid., p. 77.
35. Rigsadvokaten, act no. RA-2006-41-0151, 15 March 2006.
36. The two main sides of the debate are vividly recounted in Larsen and Seidenfaden, whose perspective is that of the liberal newspaper *Politiken*, of which the latter author is the chief editor, and in John Hansen and Kim Hundevadt, *Provoen og profeten: Muhammadkrisen bag kulisserne*, Copenhagen: Jyllands-Posten Forlag, 2006, an account from two journalists in *Jyllands-Posten*. Ironically, the two newspapers are owned by the same company.
37. The legal and administrative distinctions are subtle, and Lisbet Christoffersen makes the important point that, in fact, it is the broader sense of being acknowledged, rather than the finer points of administrative law, which matter, in her 'Anerkendt som trossamfund?', *Politiken*, 21 February 2006.
38. The incident was reported by *Politiken* on 4 February 2006.
39. Tariq Ramadan, in an interview with Lars Ellegaard and Christine Cordsen of *Politiken*, placed part of the blame for the violent responses on the delegation, and had warned them against the possibility during a visit to Copenhagen in November 2005, *Politiken*, 28 February 2006, accessed 1 March 2006 on [http://politiken.dk/].
40. Ulla Holm, *op.cit.*
41. *Politiken*, 18 September 2006.

Bibliography

Allen, C. and Nielsen, J. S., *Summary Report on Islamophobia in the EU after 11 September 2001*, Vienna: European Monitoring Centre on Racism and Xenophobia, 2002.

The Constitutional Act of Denmark of June 5, 1953, Copenhagen: Folketinget, 1999.

European Convention on Human Rights: collected texts, Strasbourg: Council of Europe Press, 1994.

Ovey, Clare and White, Robin (eds), *Jacobs and White, The European Convention on Human Rights*, 3rd ed., Oxford: Oxford University Press, 2002.

Dutertre, Gilles, *Key case-law extracts: European Court of Human Rights*, Strasbourg: Council of Europe Publishing, 2003.

DENIS SIEFFERT

The Representation of Middle East Conflicts in French Media

The conflicts in the Middle East, and especially the Israel-Palestine conflict, constitute an issue that can hardly be classified under the label 'foreign affairs' in Western countries. In France, in particular, the existence of a strong Jewish community that is never indifferent to the fate of Israel, and a Maghrebian community that shares with Palestinians this feeling of Arab identity, so admirably described by Maxime Rodinson, make it so that, for better or worse, this remote conflict never passes unnoticed. The representation of the conflict in the media is at the heart of incessant debates. My work with Joss Dray on this subject has led to a small publication entitled *The Israeli Information War*. It centred on the representation of the Israel-Palestine conflict during the period starting with the Camp David negotiations in July 2000 and ending in spring 2002 with Operation Iron Fist. Today, in order to assess journalistic behaviour and choices, one has to go further. Indeed, the anti-Palestinian propaganda is no longer – not at least exclusively – played out in the choice of words. Numerous journalists are no longer simply relays, duped or consciously, of skilful communications; they are themselves often deeply engaged in an ideological battle, convinced of its necessity. In order to avoid misunderstandings, we need to specify whom we are talking about. Of concern are mainly opinion makers; editors, columnists, commentators and analysts established in the headquarters of their media. We exclude, therefore, mostly the special envoys and, more so, the permanent foreign correspondents. Sensitive to the realities on the ground that they cannot ignore, the latter generally include a complexity in their reports that is often absent in editorials. The problem lies in the fact that the columnists, with exception of radio journalists, are in a position of extreme visibility. Their words weigh more than those of more informed correspondents.

Let us first turn back nine years. Camp David remains stencilled in our memories through the words chosen by Shlomo Ben Ami and Ehud Barak: 'the generous offer', 'the division of Jerusalem' (which is reappearing in 2007 in the run-up to the Annapolis conference), and, finally, the famous 'we no longer have a partner for peace on the Palestinian side', the formula that ushered in the Sharon-Olmert era. Today the words still retain their impact, but the choice of words attests less to a cynical propaganda effort than to something deeper, related to cultural representations. It is less the words that create the representation but rather an often subconscious – but already established – representation that determines the words. The evil is more deeply rooted. But, before even turning to the evolution of the situation and its influence on the perspective of the journalist, I would like to dwell on the pitfalls of a certain vocabulary that is specific to the media, even if the importance of this aspect needs to be put into perspective nowadays. The few words I choose here will above all help me to shed light on the functioning of the media system itself.

Retaliation

There is one word that is persistent; it is the term 'retaliation'. All Israeli violence, all bombardments, tank incursions, are always defined as 'retaliatory' action. It is thus suggested without spelling it out that violence is always initiated by Palestinians. The bombardments of Gaza, the collective punishments, come always in the form of 'retaliations' to rockets launched on Sderot. The idea does not seem to occur that no rockets would be launched if Gaza had a port, an airport, integrated economic activity, and if it wasn't the open-air prison that it is today.

On a deeper level, this accounts for a journalistic relationship to temporality that corresponds neither to history nor to the lived experiences of people. The cause is always that which just happened, yesterday. There is a quasi-professional refusal to consider more remote and deeper causes. We display a chain of events caught in a cycle of aggression and retaliation that excludes all sociological facts and simply everything from the life of the people under occupation.

Wall

There is a debate regarding the semantics of the definition of the Israeli wall in the Palestinian territories. Frequently, the two principal TV stations, TF1 and France 2, evoke the 'security barrier', enacting thus the Israeli propaganda. Others talk about the 'wall of separation', or simply about the 'wall'. As an example, the eight o'clock news aired on France 2 on 25 July 2003 displays the ambivalence of the journalists rather well. The subject appears on the evening of a meeting between George W. Bush and Mahmoud Abbas in Washington. The American president, with the sobriety characteristic of his thinking, says, 'The wall is a problem.'

The news anchor opens the subject with these words: 'This wall is erected to prevent the Palestinians from accessing Israel, but it traverses the Palestinian territories.' In the report the journalist – the correspondent in Israel – first uses the expression 'security barrier', but subsequently shows with the support of a map its trajectory 'exclusively', he says, in Palestinian territory. This example is in some ways a counterexample that proves that one mustn't focus on a single word. Overall the viewer was able to get a very clear vision of the political function of the wall. And that is essential. But the real problem resides in the exceptionality of this kind of coverage that depends on the capacities of the correspondent, and the mechanisms that order such a report. The report has not been programmed simply because there is a wall being built on Palestinian land, but because the US president speaks about it. The event is not the wall but that George W. Bush speaks of it. This is a type of, if I may say so, positive censorship. Nothing is forbidden, unlike what exists in authoritarian regimes, but the motives which permit reports of the information are extremely selective.

Colonisation[1]

One can make the same reflections about the gaping absence of information in our audio-visual media about *'colonisation'*. When researching the occurrences of the word in TV news in the last years, leaving aside the month of August 2005, time of the dismantling of the Gaza settlements, it becomes clear that there are only reports about *'colonisation'* when Ariel Sharon or George Bush decide to talk about it. The second significant occurrence on France 2 is dated 4 June 2003, the evening of the Aqaba summit. In his discourse Sharon had promised, I quote, to 'dismantle the illegal settlements'. Again one has to be careful to do justice to France 2 and to Charles Enderlin, for twenty years the permanent correspondent in Jerusalem for the public station, to have been the only one to clear the confusion between 'settlements' and 'illegal settlements' kept up by Ariel Sharon. Again a map was shown. It depicted with points of different colours the settlements, which, it needs to be remembered, are all illegal under international law, and those settlements that are called 'illegal' and which have been built without the authorisation of the Israeli government, but which are nevertheless under the state's protection. The journalist also reminded the audience that the numbers 'increased from 160,000 settlers in 1996 to 225,000 in 2003, or by 10,000 a year.'

If, however, the analysis focuses on the editorials about the conflict, often written from Paris, it is above all absence that characterises the question of the settlements in the French media. Pointing this out evokes the objection that settlement activity is not an event in itself. If criticism is directed at audio-visual media, the response will be that it is, in large measure, a reality without images that has the inconvenience of happening over long periods of time. It is here that the format of the information, or its necessarily event-focused nature, flies to the rescue of an ideology that is clearly colonial. *Colonisation* has no date. The Aqaba summit has a precise date. It is 'the tyranny of the present' that operates. Journalism is literally without history. If the Palestinians execute an operation or an attack to resist a process of expropriation and destruction of their agricultural lands or their homes, they create the event that fits the information format. The processes that provoke their wrath, though, do not find a place there. This allows me to formulate an initial answer to a question with which I am often confronted: are journalists conscious of the distortions of the reality that comes with the usage of tendentious words? Part of the answer is found in the aforementioned discussion. Often they are not conscious because they react to imperatives that present themselves as technical and that force them to show the event of the day without establishing the causal links with a history that is not, or is no longer, visible. Or, to put it differently, representation is not their problem and certainly not their 'job'.

The same mechanism was at work in November 2007. In the run-up to the Annapolis, Maryland, meeting sponsored by George Bush, *colonisation* was mentioned again because, during the preparation for the meeting, the Israelis announced plans for massive construction in the zone called E1 that risked cutting the West Bank in two. Again, it is not the act of settlement in itself that legitimises a news article but its simultaneity with the Annapolis conference. The terms 'colonisation' and 'settlement' are rare in our media. The conflict is not represented as a colonial conflict. Notwithstanding, it is exactly that in the Palestinian reality.

Terrorist

One of the few conscious debates that occur in the newsroom regards the word 'terrorist'. Many journalists have resolved this question – particularly agency journalists – by using neither the word 'terrorist', nor the phrase 'resistance fighter', suggested on the Palestinian side, but by using the word 'activist'. However, this word suffers from a pronounced ambiguity. 'Activist' is not only a euphemism that replaces the word 'terrorist', nor only a possible compromise, it is also a word frequently used by the Israeli leadership in opposition to 'civilian'. It denotes a militant or a person assumed to be a militant. It denotes the victim of a targeted Israeli assassination, someone to whom the Israelis have attributed terrorist intentions, claims that are obviously never verifiable. How often does one read news like this: 'Five Palestinians, three of them activists, have been killed in a raid by the Israeli Defence Forces'? The activist here is sentenced to death without having gone through any juridical process; nevertheless, his death appears to be almost legitimate. With disregard to the rule of law, he is sentenced and executed on the basis of intentions he is claimed to have had. The Western media have completely integrated his status of 'intentional terrorist'.

Finally, it needs to be remembered that 'terrorist' is, since the events of 11 September 2001, no longer only a descriptive term, nor a moral qualifier designating a certain type of act, notably against civilians; it is above all a political categorisation. Hamas is a terrorist organisation. Hizbullah is a terrorist organisation from the American and Israeli point of view, but not from the European point of view. This categorisation, of American origin, puts a political taboo on those it designates. It is in itself a technique of obstruction. It is rather seldom that journalists discuss its subjective character. Similarly, it is practically impossible to find from the pen of a Western journalist the designation and qualifier 'terrorist' for an Israeli air raid, which is more indiscriminate and more murderous for civilians than a suicide attack.

Islamist

To finish with this far-from-complete evocation of some of the most problematic words, I want to cite one that will allow me to link to the second part of my chapter. It is the word 'Islamist'. This one, specifically, is omnipresent in our media. Hamas is 'Islamist'. Lebanese Hizbullah is 'Islamist'. In France, the leftist intellectual Tariq Ramadan is 'Islamist'. The Iranian regime is 'Islamist'. In Pakistan, the 'Islamists' confront the army in the surroundings of the Red Mosque. The same 'Islamists' sowed death among the followers of Benazir Bhutto. Bin Ladin is 'Islamist'. In Algeria, the GIA is 'Islamist', and the Algerian branch of al-Qaʿida, which threatens attacks in France, is 'Islamist'. Young people who operate a mosque in the basement of a building in a Parisian suburb are 'Islamist'. This enumeration could go on for ages. One should, at least, worry that one single word can describe realities that are so different. But it is exactly this word that allows analysis to be tilted from an in-depth examination of a local reality, to the refusal of such an examination to the benefit of global cohesion. It is through an analysis of the ways this word is used in the Western media that one can understand what was alluded to in the introduction. The battle over the words aiming to describe a local reality is today partially outdated by the fact that there will not be any 'local realities' anymore, but a globalisation of the conflict. Thus numerous Western journalists have entered, culturally, the 'clash of civilisation' that they nevertheless zealously denounce. I refer here to the more or less conscious semantic slippage Islam-Islamist, or even Islam-Islamist-terrorist. That which is not said is often suggested in the form of suspicions. In France, the media lynching campaign that targeted the Muslim intellectual Tariq Ramadan was enlightening in this regard. Unjustly accused of anti-Semitism, this grandchild of the founder of the Muslim Brotherhood in Egypt was banned from conference halls and forbidden to participate in debates. Those among the French alter-globalisation movement that had invited him to participate in the World Social Forum in Paris in 2003 were ordered to retract their invitation, which they did not agree to do. The accusations were never frank. Many of our colleagues have used and abused a particularly insidious participle regarding this subject: the 'controversial' or 'very controversial' Tariq Ramadan. Those that created the controversy used it, therefore, to morally discredit a man who was no longer in a position to defend himself. One counters accusations gladly, but it is more difficult to counteract being the subject of a controversy. If one wants to distinguish the role of the conscious and unconscious in the use of the journalistic words, it is very clear that one reaches another level of response to the one proposed above. It is no longer merely an automatic recourse to a set of words that are never really questioned by their user. It is a political act that is almost conscious. The objective is to discredit while protecting oneself from counter attacks. The word 'Islamist', in its ambiguity, carries the same function. It discredits without being defamatory.

This brings us to the second level of the analysis presented here. It is not now a question of words that construct a representation of conflicts in the Near and Middle East. It is a representation that is already rooted in Western media culture, especially in France, and that has settled in the political and social consciences of those who produce information.

This representation rests on the conviction that the local conflicts are only pieces of one big puzzle. This explains why the Israeli-Palestinian conflict, with its strong local and historic characteristics, has in great parts disappeared from our media. The Hamas victory in the January 2006 elections, the rupture in Gaza in June 2007, brought the conflict into the template of global news. The paradox of this 'globalisation' of journalistic thinking is that it doesn't evoke a reaction from us as something which happens far away from us – after all the beauty of journalism is that it sensitises us to faraway realities – but that it makes us react to these remote events as something that happens on the street corner. According to the dominant ideology, the 'clash of civilisation' starts in our building, with that Arab neighbour, bearded and, therefore, suspect. He can resemble Bin Ladin, the trash version, or Ismael Haniyeh, the soft and thus perfidious version. The journalist becomes again, initially, a citizen in the throes of the fears of his country. If he employs the same words to define all realities, it is because for him the nebulous Islamist is everywhere.

From a semi-conscious actor for propaganda, the journalist has turned into the receiving side of a politico-intellectual discourse. Everything becomes entangled. Anything that does not resemble the idea of the 'same' is suspected of plotting against the republican pact. It is the defence of secularism that is exploited as an alibi against the expression of differences; the conviction that we have universal values to export and that these are threatened, prompting a return of colonial memories. One must not forget that for us, Arabs are, for the most part, Algerians who brought us the defeat of independence, from France in March 1962 after eight years of war. For parts of the French population, the former colonist, their heirs, their partisans, the wound is still open, and all conflict involving an Arab country brings to the fore a gnawing pain. For some French, Israeli repression is experienced like revenge.

This cultural amalgamation of places and eras leads to a sort of 'subliminal journalism'. In the current climate, I don't need to echo the Israeli government's propaganda campaign against the Palestinians. If I show an obviously very discomforting documentary about a mosque in the suburbs of Paris, including footage from a hidden camera and microphone, such as France 2 and Arte have shown in abundance, subliminally I say something about the Israeli-Palestinian conflict, or even about the Lebanese internal crisis. I hardly need to lean on local realities. They are no more than the transposition of what I can see at the foot of my building. The multiplication of the big reports on Islamism, the frequency of news magazines showing bearded faces projected against backgrounds of fire and blood, all this is enough to make me understand all the conflicts of the world in the blink of an eye, to tacitly legitimise the Israeli war on Lebanon and the idea of a future war against Iran. The media's bias in reporting and the simplification – to the extreme – of the conflict in Darfur are inscribed in the same logic. We should not forget that the label 'genocide' originates from the American neo-conservatives, and that it is, without doubt, less concerned with humanitarian considerations than with the intention to mobilise opinion against the regime in Khartoum which, according to George Bush, is part of the 'axis of evil'.

Finally, all the events in the world find their sole explanation in the verses of the Koran. How many documentaries, reports, intellectuals and journalists present us with a literal reading of the Koran, more Salafist than Salafis themselves? As if the Bible did not also abound with invitations to violence.

According to this point of view, the 'wall' is not even a 'security barrier' anymore; it is Poitiers, the French city where, as it says in the history books, Charles Martel stopped the Arabs in 732. According to this point of view it is normal to organise a conference on Palestine, like that in Annapolis in November 2007, without inviting the major political party that was the uncontested winner in the elections. According to this point of view it is not abnormal that Israel drubs with bombs all of South Lebanon, even though – behold – the French press did not hesitate to denounce certain 'excesses'. According to this point of view it is not surprising that opinion is conditioned towards a war against Iran.

Most certainly, in three years we have moved to a different representation of the Arab-Muslim world. In journalistic terms, this is evident less as lies or an inaccuracy of words, but rather in the silence, the absence of media exposure to account for the local causes of the conflicts and the profound adhesion of many a journalist to a template of an essentially religious reading. The overkill happens through amalgamation, as if the conflict had been deterritorialised and even dematerialised. This is also the reason why foreign correspondents, again, are more resistant to this evolution.

This shift happened in France during 2003–4 in the context of the debates regarding the prohibition of the *hijab* in schools. This is the delayed influence of neo-conservative culture of which Nicolas Sarkozy is the French avatar. In addition there is in France, rooted in our history and because of our difficulty in liberating ourselves from Christianity, an incapacity to consider religious facts in an anthropological manner; a panic-like fear of all religiosity, *a fortiori* when it coincides with a return of colonial memories or the instrumentalisation of anti-Semitism. The template for a religious or ethnic reading has replaced all explanations emphasising the social. Journalists are no longer the victims of a skilful communication campaign; they are taken over by a fear that they share in part with a society that has been conditioned into it. Some of them have the spirit of the crusades. They are implicated in a battle that subdues all imperatives to account for reality. Those that in the past were more or less passive relays of propaganda are today engaged in its diffusion. The journalist is not duped by propaganda; he is a crusader on a campaign. He uses the weapons of his profession to defend his children against the 'rising peril'. These are no longer just words at a distance, as much as they can be discussed, but a deep impregnation that legitimises, what without exaggerating can be called 'engagement'.

Notes

1. French language uses the word colonie to refer to Jewish settlements, giving the activity a critical connotation that is absent in English (note from the translator).

ZAHERA HARB

Grapes of Unity facing Grapes of Wrath – War Coverage and Objectivity

Preface

18 April 1996: 'Children, women and old people lay in dozens; beheaded and eviscerated ... One hundred and six innocent souls were crushed. I remember how drastic and horrendous the scene was. I remember the villagers of Qana shouting at the camera, waving their hands in every direction asking us to film and tell the world what the Israeli army had done to the innocents, to their loved ones.'[1]

Israel occupied South Lebanon for more than twenty-two years between 1978 and 2000.[2] In April 1996, the Israeli army launched a massive assault on Lebanon, aimed at uprooting the Lebanese resistance (mainly Hizbullah – the Party of God), causing massive destruction across South Lebanon and committing massacres, killing and injuring hundreds of Lebanese civilians.[3]

It was not until the war was over that one was able to recognise the importance of the role journalists played in bringing people together, uniting them in support of their fellow citizens in the south and in support of the resistance fighters against the Israeli occupation forces. Broadcasting live images of shattered, innocent bodies of the Qana victims brought solidarity among Lebanese people to its climax. After the war, the role journalists on the ground played was perceived as heroic in the Lebanese press and among the Lebanese people.[4]

Being 'heroic' was not inimical to the idea that the coverage was proclaimed as professional. The late prominent journalist and writer Samir Kassir[5] described the performance of Télé Liban (TL) during the April 1996 events as highly professional. He said in an article in the French magazine *L'Orient Express*, published in May 1996, that TL as a public service TV station proved to have ousted all competition with its wide and accurate coverage.[6] This 'act of professionalism' is still remembered as such years after.[7] There emerged a clear sense that the journalists were deeply proud of the 'objective coverage' delivered at the time.

Nonetheless, the most poignant question for me afterwards was, what kind of objectivity were we adhering to and how did we deal with journalistic norms of neutrality and balance? Additionally, what form of objectivity can journalists reporting war achieve when their country or nation is under threat?

Methodological and theoretical approach

This article is part of a larger qualitative study that investigates the Lebanese journalism culture and performance in relation to the Israeli forces' aggression against Lebanon and their encounters with the Lebanese resistance between 1996 and 2000. News values and objectivity are key aspects of the culture that this research is exploring. Journalists' admiration for truth and facts has been widely discussed in tandem with journalistic norms of objectivity, neutrality and balance. Lebanese journalists' interpretations of these values will be presented giving prominence to the ongoing discussions on the validity of the single understanding of objectivity that some scholars in the West advocate.[8]

The article will present part of the author's own story – an ethnographic account of TL's coverage during the 'Grapes of Wrath' operation. To aid the structure and flow of the narrative, it will consist of a self-reflexive narration and analysis of the events of sixteen days of TV coverage of major military incursions.[9] The performance of TL journalists, including my own, during this period will be presented and examined in relation to journalistic norms of objectivity, neutrality, balance and truth. As a self-reflexive fieldwork account, this article is necessarily of a reflexive nature and contains first person narration. It is a personal diary of the April

1996 events, produced by transcribing and translating prime-time TV news archive and head-notes.[10] Such narration is also aided by interviews conducted with most of the TL journalists and administrators who were directly involved in the April 1996 coverage. It is also aided by press clip archives. Drawing on the theoretical framework of journalism culture, the article will aim at outlining the news values and norms that characterised the work of the Lebanese journalists in their coverage of the April events and the impact of the historical, social and political context on the way they operated. Necessarily, the ethnographic tale of TL journalists' coverage and performance will be narrated retrospectively. Thus, it is a retrospective ethnographic study.

TL coverage: aiming at achieving solidarity

On 11 April 1996, the first day of the aggression, Israel bombarded the southern suburbs of Beirut for the first time since 1982 and at the same time launched an aerial bombardment on several positions in South Lebanon. According to Fouad Naim, chairman of TL (1992–6), and Aref al-Abed, TL's head of news (1993–7), the scale of bombardment and the 'intimidating statements' of Israeli officials to target the newly reconstructed infrastructure in Beirut and adjacent cities meant that these attacks were not like previous ones. The preparation to establish an outside broadcast unit in South Lebanon was under way.[11]

Voice of the Southerners

On the third day of the Israeli operation an outside broadcast (OB) van with an anchor-producer, the author; a reporter, Nadine Majzoub; a cameraman, Mahmoud Jalloul; a director, Safi al-Aris; and two members of the technical team was heading south to Sidon, known as *Saida* in Arabic.

According to Chairman Fouad Naim, the motives for such a decision were merely professional and the people who were chosen to carry out the coverage were also chosen for merely professional reasons.

> With an event like this, even if it had no patriotic aspects or national connotations to us as a company, I would still have taken the same decision to offer our audiences a full and detailed coverage. If we

were operating in a different country and an event like this took place I would have given it the same priority. If I was still the head of Agence France Press in the region, as I used to be, and this happened in Lebanon or even in Israel, if the Lebanese invaded Israel and I was in Tel Aviv, I would have made the same decision. The bonus for us here was that we were part of this nation. The attacks were affecting us and our own people's lives.[12]

Naim was making reference here to the newsworthiness of such events, which meets with the aspects of recency, immediacy and currency, which should be present in a news story for broadcasting or publication.[13] Thus, around four o'clock that afternoon developments of what we called the 'Israeli aggressive operation' against South Lebanon were covered live, minute by minute, from Sidon. Among the stories that we broke from Sidon that afternoon was the Mansouri massacre. A missile hit an ambulance carrying families escaping the fire towards a safer place, a few metres away from a UN checkpoint near Mansouri village. The massacre was accompanied by an Israeli announcement of their operation code name, 'Grapes of Wrath'.[14]

Details of when, how and where the massacre took place were given, accompanied with images of the ambulance victims in Najem hospital's morgue in the nearby city of Tyre, beside images of critically injured children lying in hospital, moaning and crying. The names of the victims were announced too. However, one victim remained unknown:

> The body of a newborn baby was brought to hospital in Sidon. It is said to be one of the ambulance victims, but no one has been able to tell us her full name yet, so she was registered as baby X.[15]

I remember that evening, I was full of fury, anger and sadness, but tried to be as calm as I could while presenting the story to the Lebanese audiences. Being professional at that point meant that I should reveal all aspects of the Israeli aggression against civilians and powerless villagers. Journalistic professional norms of objectivity and neutrality did not necessarily collide with my sympathetic reporting about the victims. Bringing the factual images to my audiences and pinpointing the scale of the Israeli aggressions were clear objectives to me. I was clearly on the side of the victims and it was not just that day's events that guided my position, but also a history of Israeli oppression and aggressiveness against Lebanese and Palestinian people,[16] some of which I lived through before I became a journalist. On that day, I realised I had two tasks to fulfil as a journalist with access to information, sources and audiences: to report and inform, and to help. However, in retrospect, a critic would say that our coverage was not impartial and the truth we were seeking was 'positioned', as no attempts were made (or even considered) to interview Israeli officials or spokespersons. However, Lebanese journalists were operating within a certain political and legal context – Lebanese laws prohibit contact between Lebanese citizens and Israelis, including journalists. Thus, we were using soundbites and statements they had given to international news agencies. Besides, talking to the 'enemy' was not seen as a necessity; instead, it was seen as an act of treason.

Thus, starting on 13 April 1996, for sixteen days TL became the voice of the southerners, those who stayed and those who escaped and took refuge in other parts of the country. Our airtime was open for refugees appealing for those whom they left behind or who had moved on and whose whereabouts were unknown. Hospitals, civil defence centres, aid organisations, government officials, Lebanese army leadership and resistance leaders were treating TL as a nationwide communications channel.

Throughout that day we were guiding refugees to public schools and public institutions that were prepared to receive them, mainly in Sidon and Beirut. At the same time our correspondents in Tyre and Nabatiyeh were interviewing civilians who had decided to stay despite the Israeli threats.

We set up a small radio-monitoring unit in Sidon to keep an ear on news from the Israelis' Arabic radio station and that of the Israeli proxy militia's radio station – Lahed militia. Israel's Arabic radio and the Lahed station were running communiqués, warning the inhabitants of South Lebanon to evacuate their villages within a certain time, threatening that the Israeli army would raze their villages to the ground.[17] We wanted to be aware of each of these intimidations in order to draw people's attention to them in our coverage and to be able to help them whether they decided to leave or to stay.

Accordingly, we were presenting the communiqués and then we were showing how people were not responding to Israeli pressure. We had news packages about how people were steadfast in their towns and villages, refusing to obey the Israeli army orders. Reports from Nabatiyeh and from Tyre showed the 'strong will of those inhabitants who are committed to their land and would never submit it to the enemy'.[18] We ran soundbites from the villagers saying: 'Where shall we go? We are not going to leave our houses for the Israelis. We were born here and we will die here.'

> The only thing we asked our colleague Zahera Harb to do in the first 24 hours of the coverage was to focus more on the people who decided to stay steadfast in their villages and towns rather than those who were leaving. We did not want to be accused by the Lebanese authorities of encouraging people to flee their villages. We did not want people to leave their land, but, at the same time, we felt responsible for informing residents about the Israeli warning statements.[19]

Apart from this request, there was no editorial guidance on how and what to report from the field. It was up to the producer-anchor in Sidon[20] to decide where to send camera crews. During the sixteen days of military operations the producer-anchors were guided by what was happening on the ground. The producer-anchor and the head of news used to conduct daily discussions over the phone about the running order of the prime-time news program and how to distribute airtime between the studio in Beirut and that in Sidon. All producer-anchors operating from Sidon interviewed for this study emphasised that there did not need to be any editorial guidance on what and how to report. They were aware of what rhetoric to use. They were operating with all the historical, political and legal aspects of the conflict with Israel in mind. For all of them Israel was the enemy that was aiming to destroy their country.

Similarly, Fouad Naim and Aref al-Abed asserted that none of the Lebanese officials interfered in TL's decision to conduct live coverage from South Lebanon.

> Even though I had a good relationship with the prime minister, neither he nor any of his entourage interfered politically in what we broadcast and what we did not. As I said before, we were dealing with an enemy of the nation and the scale of events was enormous. The Lebanese politicians were busy looking into what Israel was doing and into the scale of its operation.[21]

The Mansouri massacre was one example of how certain events dictated airtime priorities. Reuters reporter and camerawoman in South Lebanon, Najla Abou Jahjah was filming the ambulance going by when an Israeli helicopter hit it with two missiles. A few seconds later, she filmed a father carrying two heavily bleeding children in his arms, calling for help. Abou Jahjah managed to get to the hit vehicle, where victims were still inside. The scene was dreadful. The bodies of children, young girls and women were lying inside the smashed ambulance – some were still alive and moaning. Abou Jahjah focused her camera on a little girl full of dust and blood calling her aunt in a low weak voice. The voice of the little girl was mixed with the sound of the ambulance's windscreen wipers, still moving right and left on the dusty, broken windscreen.

Abou Jahjah's footage was broadcast unedited on TL from Sidon by 8.30 that evening when Reuters decided to release it. Those images were a good tool for uniting the Lebanese people against the Israeli military machine that had caused this calamity. Our coverage contributed towards achieving solidarity among the general public. Mervi Pantti and Jan Wieten describe how the media play an important role in managing emotions and repairing social life during tragedies. They state that 'media can, for instance, actively contribute to turning a climate of anxiety and fear into one of restored morale and unification.'[22] Our coverage helped to restore morale and unification. Something similar happened later with American news magazines' coverage of the events of September 11. Kitch has described in a study of this reporting how the lasting story of September 11 was not of terror, death and destruction, but one of courage, redemption and patriotic pride.[23] Thus, coverage transformed the negative feeling of fear into the positive feeling of courage and patriotism.

Reporting accurately

On 14 April 1996 the format of coverage was set: the air was open for the anchor in Sidon to feed the nation with any development taking place in the South, even if it was a small detail. At the same time an anchor in Beirut would be ready twenty-four hours a day for any breaking news on the political, diplomatic and military fronts in other parts of the country.

In addition, the airwaves were open for refugees who wanted to send messages to their families or to ask for aid supplies.

We felt that our responsibility towards our people was growing and the reasons were not just professional, but also national, as Nadine Majzoub explains:

> I tried to cover villages' bombardments as they were happening. I was always looking for the strong and expressive shots. Personally, I did not use emotional words, shots most of the time were more expressive. And I always thought that showing what was really happening with a good visual is better than using emotional and subjective words. I focused on one thing: a war was going on and the viewer had the right to see what was happening. Our main concern was to be accurate and communicate the true nature of the Israeli aggression.[24]

Nadine feels that what guided her performance was mere journalistic professionalism. However, in her quote she admitted focusing on certain images that would reveal 'the real aggressiveness' of the Israeli army. She was not subjective with the language she used, but with the images she picked. Nevertheless, Nadine believes that what she had done did not collide with objectivity as a journalistic norm. She tends to relate objectivity to being factual, as other reporters and producers of the April events did. One other thing that Nadine's quote revealed was the difficulty in detaching yourself as a war reporter from the context (whether geographic or cultural, social or humanitarian) in which you are reporting. This relates to the concept of contextual objectivity, which will be illustrated later in this chapter. Similarly, the difficulty of detachment, especially when civilian victims are involved, was expressed by other war reporters, like Kate Adie[25] and Martin Bell,[26] as mentioned above.

The importance of images as an effective tool to gain other parts of the nation's solidarity with the people of South Lebanon was recognised by all other reporters and producers without any direct guidance. All the material gathered by our correspondents and reporters from their daily tours and visits to villages, towns and cities would run live without editing, as soon as it arrived at the transmission centre in Sidon and then was edited and packaged into three-minute reports for prime-time news. In the case of severe human causalities or brutal destruction the images were run without editing, even on prime-time news, which might be extended from half an hour to two and sometimes three hours for this purpose. Lebanon's broadcast media at that time was still operating without any regulation, charters or codes.[27] Showing graphic images of bodies was seen as natural and necessary to reveal the true nature of the Israeli aggression – and thus shattered bodies on the screen was a norm that Lebanese broadcast journalists adopted during fifteen years of civil war. All journalists interviewed for this research believed that by broadcasting such graphic images they were communicating the real situation on the ground to the Lebanese and wider Arab public. The reality they were referring to was similar to what Durham and Singer[28] referred to in their analysis of journalists' coverage of Hurricane Katrina. They stated that the reality that journalists report in the time of crisis is a 'shared interpretation of reality within the larger social context of the news environment'.

Soon after we started our live coverage we began receiving phone calls from besieged villagers, informing us of what was happening in their villages. We took their numbers and started calling them back to clarify news from Radio Israel or Lahed Radio on the situation in their villages. They became our 'citizen journalists'. This way we were kept informed of what was happening to them and to the people who stayed in the villages. We were able to identify the kind of aid they needed and the appeals they wanted to bring up on national TV.[29]

Meanwhile, the Israeli attacks on the fourth day of the Grapes of Wrath operation were escalated, as was the resistance retaliation. The minister of information, Farid Makari, called upon all local, Arabic and international media organisations to keep on reporting and revealing what he called 'the barbaric nature of the Israeli aggression against Lebanon'.[30] He emphasised that accurately and objectively reporting what was happening could achieve this.

This position was portrayed in our coverage without our receiving any political guidance from any of the government officials. All the journalists who were part of the coverage and interviewed for this research made such assertions, as did the key figures on the TL administration board in similar interviews. Their assertions coincide with my own memory of how we were running the coverage at the time. Israel was the enemy and Hizbullah fighters were resistance fighters. News priority was given to the causes of

Israeli aggression and the resistance rockets shelling the Israeli Northern settlements were presented as retaliation for the Israeli aggression.

The Israeli side of the story was not absent from our coverage. Their military and political statements were given airtime, but were dealt with as claims or as 'hidden barbaric intentions'. We, as journalists, viewed Israel as an enemy state and that position was clear in the rhetoric we used to present and analyse Israeli statements and communiqués.

We were showing the impact of the Katyusha rockets[31] on the Israeli northern settlements, but after going through all the news of the Israeli aggression and in not more than two minutes and sometimes not more than seconds. The resistance acts were always portrayed as retaliation actions, an act of resistance, and Hizbullah was rarely mentioned by name as the perpetrator. Hizbullah was issuing its communiqués regarding the rocket shelling and the guerrilla operations against the Israeli army in South Lebanon under the auspices of the Islamic resistance.

However, in our coverage we were often using the word 'resistance' on its own more than the phrase 'Islamic resistance', giving it the national patriotic sense rather than confining it to a certain sectarian group. We were applying a form of self-censorship in order to achieve a unified society in support of the resistance actions. Lebanon as a whole, in all its sectors and parties, was a target for Israel; this was expressed extensively in Lebanese officials' statements and activities and was reported in our daily coverage.

Lebanon, which was torn apart during fifteen years of civil war, was coming together for the first time and starting to function as a unified civil society.[32] Churches were open throughout the country to receive the displaced. Similarly, Lebanese Muslim leaders made such appeals, and mosques were open to receive the refugees beside the public schools and public universities. All of this was shown and reported on TL, thus making its screen 'the link to the nation', which became one of the mottos of the coverage.[33]

The message: we are the victims

On the fifth day, the house speaker of the Lebanese Parliament, Nabih Berri, visited the southern city of Tyre,[34] challenging the Israeli threats to bomb the city if it was not evacuated. As one of the Shi'i leaders, who constitute the majority in population of Southern Lebanon, he called upon all the citizens who had left their houses and fled to Beirut to return to the South. A report on how people were surviving life in Tyre, and how much they were attached to their land, followed a report on Berri's visit. However, that did not cancel out the fact that those who decided to stay in nearby villages were under Israeli scrutiny, suffering from food and water shortages and lacking the simplest means of survival. Thus, the report on the opposition of Tyre was followed by images of a UN convoy carrying food supplies to besieged villagers near Tyre. This convoy was hit by Israeli shelling in an attempt to prevent the convoy from reaching its target. 'UNIFIL sources have told TL that the convoy will try getting to those villages the next day, after a new round of negotiations with the Israelis.'[35] The images carried the voices of the UN soldiers swearing at the Israelis (but without actually naming them) for bombing near the convoy vehicles and causing such panic among the team involved.

Meanwhile, the Lebanese prime minister – at the time – Rafiq Hariri[36] re-emphasised what he called 'the ground rules of the Lebanese position', gave full support to the resistance and raised question marks about the Israeli intentions behind the operation.

> Israel occupies our land and there is a resistance to this occupation. If we agree or we don't agree with the resistance political line or history or relationships,[37] that has nothing to do with the fact that our land is occupied. It is the occupation that created the resistance ... No prime minister of Lebanon agrees to deprive the resistance of its weapons as long as Israel is occupying our land. Israel is reinforcing Hizbullah by maintaining the occupation.[38]

'We are the victims, and we have the right to resist the occupation' – that was Lebanon's message to the world and to the international community. The footage of destruction and shattered bodies of innocent people broadcast on TV added value and credibility to the message. Nationally, victimising the nation and emphasising the right to fight oppression was a unifying feature. Fouad Naim denied having followed government guidance on highlighting the victims in TL's coverage. He said they were stories of newsworthiness and that the government might have actually benefited from TL's coverage to put its argument forward.[39] Naim, like other colleagues who asserted no direct guidance on what to report and what not to, dismissed the concept of any 'managed', 'planned' or 'systematic' attempts to manipulate ideas and attitudes.

TL: the compassion

On 17 April 1996 a TL crew was able to enter two besieged villages with one of the UN convoys. The convoy was bringing in food, milk, water and medical supplies. I was the reporter who accompanied the convoy from TL and to my surprise the besieged families in Bute al-Siyyad in the central sector recognised me and were more concerned about my safety than their own. One little detail that made me realise how much those people were connected with us and with the small screen was when they noticed that I did not have time to change my clothes for three days. They thought of this as being an act of solidarity with them and expressed their appreciation to me.

After seven days of coverage a promotion for TL's live coverage from the South carried the slogan 'TL, compassion'. It was broadcast to say that we were not detached from people's suffering and we were on the people's side. We sympathised with them and tried our best to be of help to them.

We were not neutral. Our audiences did not expect us to be neutral, but expected us to be objective and highlight the aggressions. This was a clear case of 'contextual objectivity'. According to Iskandar and Nawawy,[40] such contextualisation reveals 'a situational position, a way by which collectivism among participants within the same "context" – whether cultural, religious, political, or economic – is realised and engaged'. Iskandar and Nawawy state that 'it is precisely this contextualisation that aggravates and complicates the pursuit of "objective" coverage within the news media setting.'[41] They believe that:

> Contextualisation further confuses attempts at even-handedness and efforts to cover all sides of a story. Particularly in times of war, it is the

context within which a reporter operates that makes communication with the 'enemy' unacceptable.[42]

Iskandar and Nawawy's notion of 'contextual objectivity' explains the absence of any attempt by Lebanese journalists, including myself, to communicate with any Israeli official or spokesperson.

Qana massacre: a question of objectivity, neutrality and balance

One headline dominated the news coverage on 18 April: 'A day of massacres from Nabatiyeh to Qana'. Around two o'clock that day I was at Najem hospital in Tyre investigating the ambulance massacre for the weekly current affairs show I was co-producing and co-presenting with Zaven Kouyoumdjian, *Khamsa 'ala Sab'aa* (Five over Seven). While I was interviewing the director and owner of the hospital, Dr Najem, a nurse interrupted the interview to tell us they had received an appeal for help from the UN headquarters in Qana. Israel had bombarded the compound and the human casualties were massive. Minutes later, injured people started arriving at Najem hospital, the nearest to the compound. The bloody scene at the hospital was horrendous; I was not even able to keep count of the injured. I called the newsroom in Beirut and the news team in Sidon and fed them the story, emphasising the fact that something horrific had happened in Qana and that the bloody scenes in Najem hospital indicated a massacre. We called upon all medical units and ambulances to head to Qana to help in the rescue process. My cameraman was filming the wounded and the dead lying in the hospital's corridors non-stop. Blood was everywhere. Within a few minutes another TL cameraman arrived at the scene. I asked one camera crew to take the tape to the transmission unit in Sidon and head with the second camera crew to Qana. In less than fifteen minutes the rushed footage taken at Najem hospital was running on air just as it was shot, with no editing.

> Maybe our presence there partially assisted in spreading those appeals. We were asking every ambulance in the area to head to Qana and help with the rescue mission.[43]

When we arrived in Qana, the scene was clearly drastic. The moment the villagers saw us they started shouting at the camera, waving their hands in every direction, asking us to film and to tell the world what Israel had done to the innocents. They guided us to one of the containers where the villagers were sheltering. There, we stood in the middle of the shattered bodies of women, children and old people. We then heard only the voices of the rescuers asking us to film, to record the images and transmit them to the world. When we left the container, the foreign and local reporters were all over the place and most of them were crying, along with their cameramen. Some of the UN soldiers were weeping too. When we approached them they had one word to say: 'massacre'. The UN soldiers' reaction to what happened was emphasised in my piece to camera. It was a massacre and the reactions of those soldiers who were viewed as neutral were evidence of that.

> I just want to end up by saying, no one will forget the tears of a UN soldier who was crying today over the body of a child whom he was playing football with minutes before the attack took place ... It is a horrific massacre, those are the words of the UN soldiers serving in Qana.[44]

An hour after the mass crime took place, shocking images of the massacre were running uncut on air from Sidon. The tape we filmed in Qana was broadcast on TL without any editing. Through TL, only shortly after the massacre, the pictures of Qana were distributed to all Associated Press Television (APTV) subscribers around the world. According to Ahmad Hindawi, APTV's operations manager at the time and the regional manager of APTV Middle East later on, no one knew the scale of the massacre in Qana. When TL broadcast the images of the casualties in Najem hospital, he thought that the images were worth sending to the AP newsroom in London.

> I was at the TL centre in Telet al-Khayat [in Beirut] standing by the set linked to the satellite station in Jouret al-Balloot [north of Beirut] preparing to send TL's exclusive footage from Najem hospital to the London bureau, when TL broadcast the images of the massacre from Qana. I got TL's permission and plugged their live footage to the set connected straight to our bureau in London. The whole newsroom in London was in shock, I was told.[45]

The whole nation was in shock and we, as journalists, were at that point a crucial part of the nation. We did not need to do much or say much to express the cruelty of the aggressive attack that hit Qana. The Qana massacre added to the growing feeling among all Lebanese that the nation was under unjustified attack by a huge military force, which was already occupying 10 per cent of Lebanon.[46]

What made the national feeling grow more bitter was that Qana had been preceded on that day by another massacre. Early that morning an Israeli jet fired three to four rockets over a three-storey building in the southern city of Nabatiyeh, killing eleven members of the same family. Among them was a four-day-old baby called Nour. Images of a rescuer carrying Nour's dead body, after recovering it from underneath the rubble, crowned the images of the Qana victims.

Stories and memories of previous massacres committed by the Israeli army in Palestine, Egypt and Lebanon were brought up in our coverage. These related what has had happened in Qana and Nabatiyeh to previous Israeli practices towards Arab civilians since the establishment of Israel in 1948. Thus, TL's prime-time news on the evening of the massacre was full of historical connotations and references comparing what happened in Qana to what happened in the Deir Yassin massacre and that of Baher al-Bakar and al-Haram al-Ibrahimy.[47]

Zionist militias attacked the Palestinian Deir Yassin village in April 1948 and killed what was left of its inhabitants.[48] For Arabs, Deir Yassin had since become the symbol of the 'Zionist brutality and lack of mercy'.[49] By drawing parallels between the Qana massacre and Deir Yassin, Awada brought to the surface all the negative perceptions of Israel that most Lebanese and Arabs carry in their collective memory.

I was the reporter who covered the Qana massacre and Zaven Kouyoumdjian was the anchor-producer that day in Sidon, and we were thus the team that broke the story to the nation. We both used sentimental words. He was asking me about the children of Qana, about the eyewitnesses and the UN soldiers who broke down in tears. I was talking about those whom I called 'Those who suffered the survival, those who lost most of their family members and wished they were among the dead'. The images of the massacres were repeated over and over again, and stories of survival and the situations of the injured were revealed minute by minute. We were emotionally involved and we wanted them to be involved too.

> Lebanese were dying and I was covering it. It is very normal and human to be affected by such horrific scenes. I was involved. Using sentimental words or sentences were part of the 24-hour live coverage 'show' ... I think the whole coverage was a very sentimental moment in post-war Lebanon ... I was under fire. Lebanon was being attacked. I was Lebanese, reporting to a Lebanese audience on a Lebanese channel. It was very clear to me. In such moments, it is not a question of objectivity and neutrality. It is what you feel and what the nation feels. I was angry at what Israel was doing – and I showed this through my work.[50]

Zaven's questioning the need of objectivity on occasions when a journalist is an eyewitness of acts of injustice and war crimes meets with what Kate Adie[51] and Martin Bell,[52] BBC veteran war reporters, expressed in their writings.

David Mannion, editor-in-chief of ITV, also questioned the ability of reporters on the ground to be balanced and detached during their coverage of the latest Israel-Lebanon war in July 2006. Mannion told *Broadcast* magazine on 4 August 2006 that he tells his reporters to strive to be accurate, fair and honest. Mannion added:

> As for balance, it's my belief that it would quite impossible to ask an individual reporter covering one incident on one day in one war to be always balanced. Imagine this. A reporter is standing in the middle of a hospital ward. Children, some with arms and legs blown away, others blinded by shrapnel, are screaming in fear and agony. It is tough, but it is still necessary in my view to be fair and honest, but it is not reasonable and/or necessary in my view to ask him or her to be balanced. It is even okay to be angry.[53]

If being balanced is a dilemma for foreign reporters covering a war far away from home, imagine what it is like to be a reporter like TL reporters (including myself) who were reporting a war taking place in their own land and affecting their own people. One hundred and six civilians died in Qana and more than two hundred were injured, including four UN soldiers. Qana's footage dominated the screens of all local TV stations. They all had reporters in there at different times. Nevertheless, TL had the privilege of operating live from the South while the others had to drive back to Beirut with their material. Lebanese TV stations from different political and religious backgrounds were unified in the language they used to address the massacres. Anyone who died in the massacres was called a 'martyr'[54] and what Israel did was described as hostile, aggressive and a war crime against humanity.

Fouad Naim, Aref al-Abed and the ten reporters and producer-anchors who operated from South Lebanon all assured their people that they were not exaggerating or inventing stories in their coverage. For Naim it was 'the propaganda of the truth'.

> We were not hiding anything. On some other occasions we emphasised and highlighted the aggressive nature of the attacks because it was happening on our land and affecting our own people. But with Qana what the Israelis were doing did not need exaggeration. I believe what we did was reflect the reality of the attack. We were showing the images of destruction and massacres as they were. In covering the Qana massacre, I believe we were completely objective; we did not need to decorate the coverage with any national slogan or metaphor, because the images were talking for themselves.[55]

So there was a crafting of messages for a purpose – a kind of 'information war', but the aim here was to communicate what was demonstrably the truth of the Israeli aggression. As mentioned before, Naim and Abed denied any interference from the government, the prime minister, or his communication advisor in editorial decisions. TL journalists interviewed for this research assured us of this and confirmed that no one was telling them what to say or what not to say. The only guidance they were receiving was how to prioritise events. All were guided by their cultural, social and political backgrounds, in addition to their professional expertise.[56]

> We felt under attack. It was a war targeting our families, relatives and friends. We were feeling the urge to defend them. Being a citizen of this country meant that we as journalists and the people of South Lebanon were in one battle, the battle to defend our nation. One thing was clear to us; you could not be neutral in your feelings towards your nation.[57]

Patriotism and nationalistic attitudes ran high among all Lebanese journalists in the wake of the Qana massacre. Diana Moukaled of Future TV (owned by Prime Minister Rafiq Hariri), who was among the first to get to Qana, revealed that when she returned to Beirut she was completely devastated by what she had seen, such that she had almost collapsed. The prime minister's media and political advisor at the time, Nouhad al-Mashnouk, came to Future TV and calmed her down.

> He said: 'Today we won the war. We did not lose it. Those who died gave us victory. We will not let it go,' and that was what happened. Lebanon took advantage of the massacre politically on all three levels: locally, regionally and internationally.[58]

Moukaled, like the TL reporters, spoke of the relationship that connected her to the people of South Lebanon. She said that she knew what to say and what terminology to use without being guided. She grew up knowing the tragedies the Israeli occupation was imposing on the people of South Lebanon. This 'living tragedy', as she called it, was what had guided her through her work.

The Lebanese ambassador to Washington at the time, Riyad Tabbara, revealed that during the first seven days of the Israeli offence on Lebanon the American administration did not want to interfere and pressure Israel to stop its assault on Lebanon. Nevertheless, the American administration's position changed after footage of the Qana massacre dominated the screens of most international TV stations. According to Tabbara the images of Qana made them want to act quickly to negotiate a ceasefire.[59]

Conclusion

The memories encapsulated in this article brought many tears to my eyes, as if the stories were taking place now and not ten years ago. Memories of the emotional distress and frustration I felt in witnessing the killing of innocent children reminded me of how those emotions changed into anger. For sixteen days we tried to be the victims' voice in the world, or, as we felt at that time, to whoever was watching.

I realised that objectivity as a journalistic norm, and especially in covering wars, could hardly be an absolute measure of conduct. The way we, the Télé Liban journalists, interpreted objectivity identifies completely with the way we were reporting, even though we were emotionally involved with what was happening. Our 'objectivity' fell within the cultural, national and social context in which we were operating.

What the narrative or the story of this article suggests, is that journalists cannot detach themselves when covering a war launched on their country. Patriotism and nationalism thus mark the Lebanese journalists' work during these sixteen days of coverage. They identified themselves as members of the nation; they sympathised with their fellow citizens and were emotionally involved.

Partisanship in TL journalists' commentaries became clearer in the wake of the Qana massacre. The chairman of TL, Fouad Naim, was the only one who answered the question of whether our coverage might be labelled as propaganda by saying, 'If anyone accuses us that we were conducting propaganda, then yes we were. It was propaganda of the truth.' Looking at it retrospectively and from a distance, it was surely a 'positioned truth', a 'truth' told through the eyes of one side of the conflict, but none of us saw it as such at that moment. It was our version of the truth, which then – because it was believed in so fervently and so nationalistically, because it was lived and embodied in the daily encounters with shattered and innocent bodies – comes to seem to be an objective truth.

We were all driven by our experiences as Lebanese citizens over the course of a lifetime, by our sense of the wrong being done to our country. We were driven by the need to report and highlight the scale of the Israeli offence and to communicate the suffering of the residents of South Lebanon to the rest of the Lebanese public.

There was no orchestrated or managed campaign. Journalists were acting upon their own beliefs and ideas. Israel was an enemy state that was occupying their land and was launching an attack on their civilian people. Their ideology was simply generated from the history of the Arab-Israel conflict and the Lebanese-Israeli part in this. The purpose was to unify the people against one enemy, Israel, and stay defiant.

Thus, the shared ideas and ideals, the shared experiences, the shared threat, the interpersonal relationships, produced a very consistent and common approach, which became very deliberate in terms of personal commitment. So the way we reported was intentional, though the media campaign was not managed or planned.

Our coverage was factual. We did not fabricate news; we did not deceive our audiences. The footage of guiltless victims killed by Israeli bombardments and air raids spoke for itself. We all believed that we were objective in our reporting. However, as discussed earlier, ultimate objectivity was actually unachievable. Journalists are affected by the historical, cultural, social and political context they operate within. Thus, the objectivity all journalists interviewed for this research claimed is contextual. Journalists reporting war when their country or nation is under threat can hardly achieve any other kind of objectivity than precisely this.

The Lebanese journalists believed they were serving their people's cause by highlighting and exposing the Israeli aggression. I was personally affected by the scale of the Israeli brutality in killing civilians and destroying homes, and felt compelled to report this Israeli brutality to my audiences.

However, looking back at my performance in a retrospective and reflexive way, I realise that my reporting was affected by the horror I had lived through and witnessed at the time. As a journalist I 'went native', in the anthropological and ethnographic sense, and, as anthropologists have come to understand, objectivity and distance are impossible in such a context, where the only truth is that which is positioned or contextual. Impartiality in such circumstances is more than difficult to maintain. But these complexities are exactly what were involved in what the TL journalists were doing in this context.

Notes

1. Harb, Zahera in BBC News, 31 July 2006. These strong memories have remained present, even years after the massacre took place. It became part of my collective memory as a Lebanese citizen and as a journalist.
2. On the 25 May 2000 Israeli troops completed their withdrawal from the majority of the occupied territories in South Lebanon, keeping a small piece of disputed land called Shebaa Farms. The area is located at the junction of Syria, Lebanon and Israel. It is 14 km (9 miles) in length and 2.5 km (2 miles) in width. Israel claims it is Syrian land and Lebanon says it is Lebanese land. Syria says it is Lebanese, but has not supplied the United Nations with any written documents on the issue yet (see BBC News, 25 May 2000 and Chassay, *The Guardian*, 10 October 2006).
3. For sixteen days I was positioned in South Lebanon reporting the Israeli assaults on the villages and cities of the south for Lebanese state-run TV, Télé Liban.
4. By reporting and informing the public of what was really happening in South Lebanon, the journalists on the ground had fulfilled their duty.

Nationalistic and patriotic reporting was seen as part of the role of a professional journalist. The patriotic and nationalistic approach made the coverage of the events in the South 'heroic'.

5. Samir Kassir was assassinated on 2 June 2005 in Beirut, a bomb having been planted under his car seat. He was one of Independence 05 movement in Lebanon's political architecture. The movement followed the assassination of Prime Minister Rafiq Hariri on 14 February 2005.
6. Kassir praised what he called the 'calmness and professionalism' of my performance specifically, which in his words 'projected the voice of an emblematic figure of the coverage of TL' (Kassir, *L'Orient Express*, May 1996).
7. Roula Abdallah, of the daily *al-Mustaqbal*, wrote on 29 March 2007: 'Zahera and her colleagues raised the audiences' confidence in the Lebanese media to inform, [and] without the need to quote international agencies like Reuters or AFP' (Abdallah, *al-Mustaqbal*, 29 March2007: 8).
8. Campbell, *Information Age Journalism*.
9. The narration is based on forty hours of TL prime-time news between 11 April 1996 and 27 April 1996.
10. Memories brought up by watching the news coverage – see Sanjek, 1990: 93.
11. Al-Abed, interview with the author, 2004.
12. Naim, interview with the author, 2004.
13. Roscho in Tumber, *News: A Reader*.
14. The Israeli government code-named its operation after John Steinbeck's famous novel, *The Grapes of Wrath*, relating rhetorically its operation with the theme of fighting for existence. (Ashi, TL news, 19.30, 13 April 1996).
15. Harb, TL news, 19.30, 13 April 1996.
16. See Cooley, *Green March, Black September*; Friedman, *From Beirut to Jerusalem*; Fisk, *Pity the Nation: Lebanon at War*; and Picard, *Lebanon a Shattered Country*.
17. Communiqué No. 4, Lahed Radio, 13 April 1996.
18. TL news, 19.30, 13 April 1996.
19. Al-Abed, interview with author, 2004.
20. TL anchor-producers that covered the 'Grapes of Wrath' operation from South Lebanon were:
    Zahera Harb on 13/14/15 April 1996; 23/24/25/ April 1996, and 27 April 1996;
    Zaven Kouyoumdjian on 16/17/18 April 1996;
    Dalal Kandeel on 19/20 April96 and 26 April 1996;
    Rajaa Kamouneh on 21/22/ April 1996.
    Head of news (1993–1997) Aref al-Abed described the teams that covered

the events from South Lebanon as suicidal. He praised most the first team that agreed to go south at a time where others were reluctant and were questioning the benefits of doing so (al-Abed, interview with author, 2004).
21. Naim, interview with author, 2004.
22. Pantti and Wieten, 'Mourning Becomes the Nation: Television Coverage of the Murder of Pim Fortuyn', p. 304.
23. Kitch, 'Mourning in America: Rituals, Redemption, and Recovery in News Narrative after September 11', p. 222.
24. Nadine Majzoub, interview with author, 2004.
25. See Adie, K., 'Despatches from the Front: Reporting War', *Contemporary Issues in British Journalism*.
26. See Bell, M., 'The Journalism of Attachment', in Kieran, M., *Media Ethics*.
27. See Boulus, *Television History and Stories* and Dajani, 'The Changing Scene of Lebanese Television'.
28. ICA conference, 2006: 6.
29. TL news, 19.30, 14 April 1996.
30. Makari, TL news, 19.30, 14 April 1996.
31. Soviet-made, multiple rocket launchers with short-range rockets.
32. Iskandar, *Rafiq Hariri and the Fate of Lebanon*, p. 73.
33. TL news, 19.30, 14 April 1996.
34. Called *Sour* in Arabic, 83 kilometres south west of Beirut, the fourth largest city in Lebanon.
35. Harb, TL news, 19.30, 15 April 1996.
36. Former Lebanese Prime Minister Rafiq Hariri was assassinated in a car bomb in Beirut on 14 February 2005. Iskandar explains in his book *Rafiq Hariri and the Fate of Lebanon* (2006), how Hariri invested his personal relations with the world leaders to support the Lebanese cause.

    Hariri secured [French President Jacques] Chirac's undiluted help in responding to the tragedy, convinced [British prime minister at the time] John Major of the enormity of the butchery, persuaded [the then Russian president] Boris Yeltsin to champion the Lebanese cause and to chastise Israel at least verbally, and passed a convincing message to the German chancellor. In the United States, Hariri depended on the support of Prince Bandar Bin Sultan, the Saudi ambassador who had developed close ties with President Clinton, as he had with George Bush Sr. Clinton was fearful of the loss of his Middle East peace initiative and therefore instructed his secretary of state, Warren Christopher, to do everything possible to address the impact of the Qana slaughter (Iskandar, 2006: 80).
37. Referring to Hizbullah relations with Iran. For more information on the link between Hizbullah and Iran see Hamzeh (2005).

38. Hariri in TL news, 19.30, 15 April 1996.
39. Naim, interview with author, 2004.
40. In Allan and Zelizer, *Reporting War Journalism in Wartime*, p. 320.
41. Ibid.
42. Ibid.
43. Harb, TL news, 19.30, 18 April 1996.
44. Harb, TL news, 19.30, 18 April 1996.
45. Hindawi, interview with author, 2004.
46. See *al-Safir*, 19 April 1996.
47. Massacres committed by the Israeli army against the Palestinians in 1948, 1956 and 1992. Awada, TL news, 19.30, 18 April 1996.
48. See Makovsky,
49. Ibid.
50. Kouyoumdjian, interview with author, 2004.
51. See Adie, K., 'Despatches from the Front: Reporting War', *Contemporary Issues in British Journalism*, and Adie, K., *The Kindness of Strangers*.
52. Bell, M., *Through Gates of Fire*.
53. Mannion, *Broadcast*, 4 August 2006.
54. In Lebanon, the word martyr is also used to refer to innocent civilians who die from an aggressive act and thus not just to fighters. The term has been used during the last thirty-two years to refer to persons who died in conditions of oppression, no matter to what religious sect they belong.
55. Naim, interview with the author, 2004.
56. Kandeel-Yaghi, interview with author, 2004.
57. Kamouneh, interview with author, 2004.
58. Moukaled, interview with author, 2004.
59. Interview with author, 1999.

Bibliography

Adie, K., *The Kindness of Strangers*, London: Headline, 2002.

Adie, K., 'Dispatches from the front: reporting war', *Contemporary Issues in British Journalism*, Cardiff University: Vauxhall Lectures, Centre for Journalism Studies, 1998.

Al-Abed, A., *Lebanon and Al Taef*, Beirut: Centre for Arab Unity Studies, 2001.

Allan, S. and Zelizer, B., *Reporting War Journalism in Wartime*, London: Routledge, 2004.

Alterman, J., *New Media New Politics? From Satellite Television to the Internet in the Arab World*, Washington: Washington Institute for Near East Policy, 1998.

Atkinson, P., *Understanding Ethnographic Texts*, USA: Sage University Paper, 1992.

Aull Davies, C., *Reflexive Ethnography*, London: Routledge, 1999.

Bell, M., *Through Gates of Fire*, London: W&N, 2003.
Bell, M., 'The Journalism of Attachment', in Kieran, M., *Media Ethics*, London: Routledge, 1998, pp.15–22.
Blanford, N., *Killing Mr Lebanon: the Assassination of Rafik Hariri and its Impact on the Middle East*, London: I. B. Tauris, 2006.
Boulus, J., *Television History and Stories*, Lebanon: FMA, 1995.
Bowen, J., *War Stories*, London: Simon & Schuster, 2006.
Boyd, D., *Broadcasting in the Arab World*, USA: Iowa State University Press, 1993.
Campell, V., *Information Age Journalism*, Great Britain: Arnold, 2004.
Cooley, J., *Green March, Black September*, London: Frank Cass, 1973.
Dajani, N., 'The Changing Scene of Lebanese Television', *TBS Electronic Journal*, Fall/ Winter 2001, no. 7, [http://www.tbsjournal.com/Archives/Fall01/dajani.html. (Accessed: 09.01.05)].
Durham. F., & Singer, J., *The Watchdog's Bark: Professional Norms and Institutional Routines in Cable News Coverage of Hurricane Katrina*, paper presented at the International Communication Association (ICA) Conference, 19–23 June, Dresden Germany, 2006.
El-Nawawy, M. & Iskandar A., *Al Jazeera: How the Free Arab News Network Scooped the World and Changed the Middle East*, USA: Westview, 2002.
El-Nawawy, M. & Iskandar A., *Al-Jazeera: The Story of the Network that is Rattling Governments and Redefining Modern Journalism*, USA: Westview, 2003.
Firmo-Fontan, V., 'Power, NGOs and Television', in Sakr, N., *Women and Media in the Middle East*, London: I. B. Tauris, 2004.
Fisk, R., *Pity the Nation: Lebanon at War*, Oxford: Oxford University Press, 2001.
Fisk, R., *The Great War for Civilisation: the Conquest of the Middle East*, London: Harper Perennial, 2006.
Friedman, T., *From Beirut to Jerusalem*, London: Harper Collins, 1998.
Hall, J., *Online Journalism*, London: Pluto Press, 2001.
Iskander, M., *Rafiq Hariri and the Fate of Lebanon*, London: Saqi books, 2006.
Kean, F., *All of Those People*, London: Harper Perennial, 2005.
Kitch, C., 'Mourning in America: Rituals, Redemption, and Recovery in News Narrative after September 11', *Journalism Studies*, no. 4 (2), 2003, pp. 213–244.
Palmer Harik, J., *Hezbollah: The Changing Face of Terrorism*, London: I. B. Tauris, 2004.
Pantti M., and Wieten J., 'Mourning Becomes the Nation: Television Coverage of the Murder of Pim Fortuyn', *Journalism Studies Journal*, no. 6 (3), 2005, pp. 301–313.
Pantti, M., 'Masculine Tears, Feminine Tears – and Crocodile Tears: Mourning Olof Palme and Anna Lindh in Finnish Newspapers', *Journalism Journal*, London: Sage, no. 6 (3), 2005, pp. 357–377.

Pavlik, J., *Journalism and New Media*, New York: Columbia University Press, 2001.
Picard, E., *Lebanon a Shattered Country*, New York: Holmes & Meier, 2002.
Sanjek, R., *Fieldnotes*, London: Cornell University Press, 1990.
Simpson, J., *Strange Places, Questionable People*, London: Pan Books, 1999.
Simpson, J., *The Wars Against Saddam: Taking the Hard Road to Baghdad*, London: Macmillan, 2003.
Tumber, H., *News: A reader*, New York: Oxford University Press, 1999.

Newspaper archives and online sources:
Abdallah, R., 'Zahera Harb, Qana's reporter in Lebanon ... and her footprints in London', *al-Mustaqbal*, 29 March 2007, p. 9.
Al-Safir (JPEG) editions 1 April 1996 to 30 April 1996, CDROM-Arab Documentation Centre, Beirut, Copy held by author.
BBC in 'Focus: Shebaa Farms', *BBC News*, 25 May 2000, [http://news.bbc.co.uk/1/hi/world/middle_east/763504.stm (accessed on 02.02.07)].
Chassay, C., 'Call for Israel to leave Shebaa Farms', *The Guardian*, 10 October 2006, [http://www.guardian.co.uk/israel/Story/0,,1892111,00.html (accessed 02.02.07)].
Harb, Z., 'Viewpoint: Remembering Qana', *BBC News*, 31 July 2006 [http://news.bbc.co.uk/2/hi/middle_east/5231832.stm (accessed 01.02.07)].
Kassir, S., 'Public service and professionalism [Service public et professionnalisme]', *L'Orient Express*, May 1996, p. 32.
Mannion, D., 'Fighting to Tell the Truth', *Broadcast Magazine*, 4 August 2006, np.
TL News Archives:
Télé Liban prime time news programs (19.30 pm)

11/04/96	Presenter from Beirut studios: Nada Saliba	
11/04/96	Presenter from Beirut studios: Tony Salameh	
13/04/96	Presenter from Beirut studios: Souad al-Ashi	
	Presenter-producer from Sidon: Zahera Harb	
14/04/96	Presenter from Beirut studios: Naeemat Aazouri	
	Presenter-producer from Sidon: Zahera Harb	
14.04.96	Prime Minister Rafiq Hariri, interview with CNN correspondent Brent Sadler (re-broadcast)	
15/04/96:	Presenter from Beirut studios: Souad al-Ashi	
	Presenter-producer from Sidon: Zahera Harb	
16/04/96	Presenter from Beirut studios: Souad al-Ashi	
	Presenter-producer from Sidon: Zaven Kouyoumdjian	
17/04/96	Presenter from Beirut studios: Neemat Aazouri	
	Presenter-producer from Sidon: Zaven Kouyoumdjian	
18/04/96	Presenter from Beirut studios: Wasef Awada	

	Presenter-producer from Sidon: Zaven Kouyoumdjian
19/04/96	Presenter from Beirut studios: Souad al-Ashi
	Presenter-producer from Sidon: Dalal Kandeel
20/04/96	Presenter from Beirut studios: Naamet Aazouri
	Presenter-producer from Sidon: Dalal Kandeel
21/04/96	Presenter from Beirut studios: Tony Salameh
	Presenter-producer from Sidon: Rajaa Kamouneh
22/04/96	Presenter from Beirut studios: Naamat Aazouri
	Presenter-producer from Sidon: Rajaa Kamouneh
23/04/96	Presenter from Beirut studios: Nada Saliba
	Presenter-producer from Sidon: Zahera Harb
24/04/96	Presenter from Beirut studios: Tony Salameh
	Presenter-producer from Sidon: Zahera Harb
25/04/96	Presenter from Beirut studios: Souad al-Ashi
	Presenter-producer from Sidon: Zahera Harb
26/04/96	Presenter from Beirut studios: Wasef Awada
	Presenter-producer from Sidon: Dalal Kandeel
27/04/96	Presenter from Beirut studios: Souad al-Ashi
	Presenter-producer from Sidon: Zahera Harb

Interviews and correspondence with the author:

Abou Jahjah, N., Reuters reporter, Télé Liban OB Transmission, Sidon, 5/7 weekly programme, 19.04.96.

Al-Abed, A., Télé Liban's Head of News; correspondence from Bahrain, 24.03.04.

Awada, W., Télé Liban news editor; his office, TL headquarters, Telet al-Khayat, Beirut, 19.04.04.

Hindawi, A., head of APTV regional operations; his office, Beirut, 04.01.04.

Kamouneh, R., TL presenter-producer; her office, Lebanese Foreign Ministry, 17.04.04.

Kandeel, D., TL presenter-producer, her house; Haret Hriek, Beirut Southern Suburb, 06.01.04.

Kandeel, N., deputy head of National Media Council; Lebanese parliament, Beirut, 15.04.04.

Koumedjian, Z., TL producer-presenter, via e-mail, Beirut, 20.04.04.

Majzoub, N., TL reporter, via e-mail, 05.04.04.

Moukalled, D., Future TV reporter; Future TV offices, Raoushe, Beirut.

Naim, F., TL chairman; via phone from Paris, 06.02.04

Saad, H., TL news director; her house, Aramoun, south of Beirut, 07.01.04.

Tabbarah, R., former Lebanese ambassador to Washington, his office Beirut, 25.03.1999.

EHAB BESSAISO

Media War or War on Media: Lessons from Iraq

There is no doubt that the 2003 Iraq War marked a new era in warfare in which war and media technologies directly affected the conduct of war. Various techniques and methods were implemented during this conflict to promote effective media strategies which could control information and manage news and images. These techniques included the embedding policy, the use of language and terminologies, the participation of PR firms and the role of media corporations, which contributed to and influenced the media's performance during the war.

My analysis of the major media strategies by the US and UK governments covers the period of time up to 9 April 2003, the date Saddam's statue was toppled, as this is argued to have marked the end of the war, and the defeat of the Iraqi army, regardless of President Bush's declaration on 1 May 2003 of the end of the major offensive in Iraq. The Iraqi regime had practically collapsed on 9 April 2003, and it is argued that everything after this date could therefore fall within the ongoing occupation period of Iraq.

*Communication strategies and wartime
propaganda during the Iraq War*

It is believed that the effectiveness of communication strategies strongly relies on the development of technology in order to efficiently shape public opinion and win its support. During the 2003 Iraq War, it was observed that major developments in communication strategy and wartime propaganda occurred through the planning and conducting of the military campaign. This is perhaps what makes that war the most sophisticated so far. As stated by Martin Bell, major changes have occurred since the 1991 Gulf War in terms of technology and the military.[1] However, it can be argued that the significance of this conflict extended beyond the remarkable use of modern technology in war reporting, or the implementation of new media techniques to manage and control information: it is the substantial integration of the media into the war machine and the effects of this on the coverage. According to Robin Brown, there are three influential communication tools used by the US 'in waging war on terrorism'. These are 'military concepts of information warfare, foreign policy concepts of public diplomacy, and approaches to media management drawn from domestic politics.'[2]

The first concept is based on the information operations (IO) doctrine which, according to Brown, can be seen as 'a systematic attempt to make sense of warfare as an exercise in information processing'. Computer systems, as described by Brown, are tools to gather, process and disseminate information so that IO can include any effort to attack or defend the information necessary for the conduct of operations. However, the military took the idea of information warfare a step further, providing a coherent framework to bring together existing activities, 'psychological operations (PSYOPS), deception, public affairs – that is the military-press interface – and civil military affairs with computer network operations.'[3]

The second concept is public diplomacy. Brown suggests it aims to draw together international broadcasting, cultural diplomacy, educational exchanges and overseas information activities.[4]

The third principle is political news management or 'spin', which Brown described in institutional terms as 'a newcomer to the international arena.' In defining the role of spin, Brown declared that it is to develop elaborate mechanisms to secure and control media coverage. This machinery normally focuses on short-term media coverage and a strategic communications function that develops proactive communications strategies, 'for instance by using the activities of leaders to communicate key messages.' Therefore, 'spinning' aims to persuade the media that one version of reality rather than another is the 'real' story and that the way the spinner has told it is the correct one. This approach would seek to balance an active approach to shaping the media environment with a broader commitment to some rules of the game.[5]

Furthermore, Justin Lewis et al stated that in an age of increasingly sophisticated digital and satellite communications, the use of information as a weapon is becoming a dominant feature of modern warfare. Information management is also an increasingly important part of global government and foreign policy.[6] It was therefore the impact of information dominance during the war on Iraq, which was the key strategy in the US and UK to influence public opinion both domestically and worldwide to support the war efforts and the governments' policies. According to David Miller, the concept of information dominance is the key to understanding the US and UK's respective propaganda strategies, as it redefines our notions of spin and propaganda and the role of the media in capitalist societies.[7] In his analysis, Miller pointed out two new elements seen in information dominance when compared to traditional conceptions of propaganda.

> The first is the integration of propaganda and psychological operations into a much wider conception of information war. The second is the integration of information war into the core of military strategy.[8]

According to Miller, traditional conceptions of propaganda involve crafting the message and distributing it via government media or independent news media. Current conceptions of information war go much further and incorporate the gathering, processing and deployment of information by way of computers, intelligence and military information systems (command and control). The key preoccupation for the military is interoperability, where information systems talk to and work with each other.

Interoperability is a result of the computer revolution, which has led to a so-called revolution in military affairs. Now propaganda and psychological operations are simply part of a larger information armoury. As Colonel Kenneth wrote, the 2003 attack on Iraq 'will be remembered as a conflict in which information took its place as a weapon of war.'[9]

Furthermore, the experience in Iraq in 2003, according to Miller, showed the development of the planned integration of the media into the instruments of combat. It also showed the increased role of the private sector in information dominance, a role that reflects wider changes in the armed services in US and UK. Miller regarded applying the concept of information dominance to media management's activities would also allow the US and UK's strategists to tolerate dissent in the media and alternative accounts on the Internet.

Dissent only matters if it interferes with their plans. As US military authors Winters and Giffin put it, 'Achieving information dominance involves two components: first, building up and protecting friendly information; and second, degrading information received by your adversary.' Both of these refer not simply to military information systems but also to propaganda and the news media.[10]

Therefore, Miller believed that the application of the embedding policy was a clear means of building up and protecting friendly information, in addition to allowing journalists better access to the fighting than that given in any conflict since Vietnam.[11] Nonetheless, Martin Bell, who regarded the embedding policy as a bold experiment which confounded many predictions, including his own 'that it would muzzle the press and deliver only news that pleased the Pentagon,' argued that it created confusion among journalists who decided to embed with the military about their role in covering the battles. 'Not only the younger embedded journalists, but some of the older hands as well, failed to grasp the difference between being *with* an army or being *of* an army.'[12]

Bill Katovsky, however, defined the embedding policy as 'a slick new public-relations concept'. For Katovsky, the introduction of this concept was due to the need for war planners, in the build-up to the new Iraqi campaign, to toss aside any lingering doubts they had about the media.[13] In agreement with Katovsky, Howard Tumber and Frank Webster stated that the military attempted a range of strategies to ensure that they got appropriate and acceptable reportage about their activities, from long-term cultivation of contacts in the media to the granting of privileged access to favoured reporters, to aggressive physical threats towards journalists who got too close to situations the military did not want them to see.[14]

Pentagon officials, according to Philip Seib, made it clear that the embedded journalists would receive better treatment than the approximately 1800 unilateral journalists. When Kuwait blocked some 'unembedded' reporters from entering Iraq, Pentagon spokesperson Bryan Whitman said:

> We are going to control the battle space. Reporters that are not embedded are going to be treated like any other civilian, approached with a certain amount of caution. For many journalists, proving their identity can sometimes be problematic.[15]

Whitman justified this policy by raising the assumption, from the angle of security, that the Iraqis might have 'individuals pose as journalists.'[16] Nevertheless, Katovsky stated that in the early days of the war, several unilateral journalists were killed by venturing too close to the fighting. Two other journalists were killed while being embedded.[17]

At this point, it would be also worth mentioning that the other impact of the use of advanced war technology during the Iraq conflict was the scope of devastation and the nature of the victims, including journalists. According to the International Press Institute (IPI), 'at least fourteen journalists were killed during the six-week campaign; with two others missing and believed dead.' However, the hostilities did not end after President Bush declared on 1 May 2003 that 'major operations in Iraq [had] ended.' The death toll was to grow to nineteen by the end of 2003.[18]

Furthermore, in a report by Reporters Without Borders (RSF) on the journalists who were killed in Iraq between 20 March 2003 and 20 March 2006, the organisation stated that the war on Iraq had proved to be the deadliest for journalists since World War 2. 'A total of eighty-six journalists and media assistants have been killed in Iraq since the war began.'[19]

This number, according to RSF, is more than the number killed in Vietnam or even the civil war in Algeria. Around sixty-three journalists were killed in Vietnam during the twenty years from 1955 to 1975 and seventy-seven journalists and media assistants were killed during the civil war in Algeria from 1993 to 1996.[20]

On the other hand, Kari Lydersen argued that one aspect of the domination of news reporting is the control and manipulation of viewpoints and of information coming directly from the government. This, according to Lydersen, was an explanation of why attacking the alternative media on certain occasions had its own strategic justifications.

> The Bush administration has also been hard at work on limiting and, ideally, silencing, opposing or challenging viewpoints and factual narratives coming from other sources. The administration has attacked

al-Jazeera, the Qatar-based and state-funded media outlet that is the primary news source for much of the Arab world.[21]

As John Simpson suggests, whether the attack on al-Jazeera's office in Baghdad was deliberate or not, it was disturbing that the channel's offices had been hit twice in two consecutive wars – in Kabul in 2001 and in Baghdad in 2003 – and that 'on each occasion American command had complained that al-Jazeera was supporting the enemy'.[22]

Similarly, the justification for bombing the Iraqi state television station on March 2003 by a joint US/UK mission, which resulted in many civilians being killed, was that the station was 'part of a command and control centre' and it was housed in 'a key telecommunications vault' for satellite communications. Yet, according to Simpson, 'John Gibson of Fox News, by contrast, claimed the credit for this attack.' Gibson stated that 'Fox's criticism about allowing Saddam Hussein to talk to his citizens and lie to them has had an effect.'[23]

Media battles and news management

As indicated by Lewis et al, the 'key messages' of the 2003 coalition's media operations in Iraq were carefully crafted and presented to the global public by teams of professionals skilled in the art of perception management, and considerable effort was put into the presentation of a united front at the more strategic level.[24] Thus, the need to structure a media campaign in order to push the message on a daily basis was a key element in the efforts of media strategists to win the public's support; therefore the White House defined the invasion as a war of liberation and produced a media campaign to support that idea.

The official theme was 'Iraq: From Fear to Freedom' and the news media were given 'liberation updates' and heard 'voices of freedom' from Iraqis who appreciated Saddam Hussein's ouster.[25]

According to Tumber and Palmer, information planning by the US government before the war was based on a twenty-four-hour news cycle, a kind of global PR network to be activated from different parts of the world, from the Pentagon, from Qatar and from the embedded journalists.[26]

Furthermore, Suzy Defrancis, President Bush's deputy assistant for communications, outlined the media relations that were to be introduced.

> When Americans wake up in the morning, they will first hear from the [Persian Gulf] region, maybe from General Tommy Franks, then later in the day, they will hear from the Pentagon, then the state department, then later on the White House will brief.[27]

In keeping with Tumber and Palmer, Ari Fleischer, the White House press secretary, used to set the day's message with an early morning conference call to Alastair Campbell, a conference call to White House Communications Director Dan Bartlett, State Department spokesperson Richard Boucher, Pentagon spokesperson Victoria Clarke, and White House's Office for Global Communications Director, Tucker Eskew. This routine was similar to the procedures introduced during the Afghanistan War in 2001.[28]

Moreover, Angus Taverner stated that the MoD was torn between fulfilling its public duties to keep the British public informed and that of keeping issues of national security closely guarded.

> We learned in Kosovo, when there were a lot of incorrect accusations made against NATO by the Milosovic regime, that the media won't wait, and the timeliness of our response is the key. The one battle we lost in the Falklands was the media battle, and it produced a lot of work in its wake. By the time we got to the last Gulf War, people thought we had come a long way but we still have not got everything right.[29]

Nonetheless, Steven Tatham, who served as the Royal Navy's public spokesman during the Iraq War, stated that while the world's media waited for the conflict to begin, both Britain and US had already begun their media-handling strategies months before. In his explanation of the efforts to manage the media battle space during the war, Tatham indicated that the US military had a dedicated uniformed public affairs organisation, which provided an entire military career structure, from private to general, for public affairs officers (PAOs).[30]

In keeping with Tatham, for most media operations officers the major support is that offered by the MoD press office and its updated versions of lines to take on particular issues, key themes and messages and question-and-answer materials. These briefing notes provide the central government's direction on what can and, importantly, what cannot be said.³¹

Information control, spin and the role of the private sector

The Pentagon's desire for the total control of information extended well beyond Iraq into the living rooms and bedrooms of America and around the world. That was, according to Paul Rutherford, driven in part by the dubious assumption about 'The CNN Effect', the way television images could impact on war decisions – for instance, the claim that 'television got us in and television got us out' in Somalia a few years ago. More important was the realisation that victory was the single most important product that had to be sold to the citizens as taxpayers and consumers.³²

In keeping with Rutherford, early in 2002 rumours hit the press that the Pentagon was playing with the notion of 'Black Propaganda', the planting of false stories, not in the US but elsewhere in the world. Indeed, Maud Beelman, the director of the International Consortium of Investigative Journalists in Washington, told CBS Radio during the war that the Pentagon was grouping its public-affairs officers and its psychological operations into one discipline called Information Operations.³³

For propaganda expert Nancy Snow, propaganda has not changed in terms of its purpose.³⁴ However, according to Randall Bytwerk, the primary change is the technology rather than the method. 'It is now possible to spread much more information, much faster.'³⁵ Therefore, various techniques were adopted and used by strategists to manipulate the media, in order to sell the war and to control the message. As Danny Schechter states, these techniques targeted the media through cultivation and co-operation. They included message development, polling, psychological warfare and 'perception management'. A number of intelligence directorates, public diplomacy channels and public relations firms were used, as were civilian and military agencies.³⁶

Moreover, as indicated by Schechter, what was additionally new in the war on Iraq was the sophisticated way in which corporate public-relations techniques were adapted by the Bush administration to create the rationale for the war, to orchestrate support for it, bring the media on board and then sell it to politicians and to the public.³⁷

According to Deepa Kumar, the mechanisms used for information control were successful due to two prime co-existing factors:

> The development and testing of government information control strategies over the last three decades, and the emergence of a for-profit conglomerate media system that lends itself to propaganda due to its structural limitations.[38]

Supporting such ideas, Douglas Kellner argued that the media have become the 'arms of conservative and corporate interests', due to the concentration of ownership. Thus, instead of acting in the interests of the public, they advance the interests of political and economic elites.[39] For that reason, Norman Solomon stated that the military-industrial complex 'extends to much of the corporate media.'[40]

Solomon pointed out that media-owning corporations can also be significant weapons merchants. In a combined study with Martin A. Lee, he discovered in 1991 the stake one major company had invested in the latest war.

> What we found was sobering: NBC's owner, General Electric, designed, manufactured or supplied parts or maintenance for nearly every major weapon system used by the US during the Gulf War – including the Patriot and Tomahawk cruise missiles, the Stealth bomber, the B-52 bomber, the AWACS plane, and the NAVSTAR spy satellite system. 'In other words,' we wrote in *Unreliable Sources*, 'when correspondents and paid consultants on NBC television praised the performance of US weapons, they were extolling equipments made by GE, the corporation that pays their salaries.'[41]

Additionally, Solomon revealed that during one year, 1989, General Electric had received close to $2 billion in military contracts related to systems that were eventually being utilised in the Gulf War.

Fifteen years later, the company still had a big stake in military spending. In 2004, when the Pentagon released its list of top military contractors for the latest fiscal war, General Electric ranked eighth with $2.8 billion in contracts.[42] According to Schechter, the benefits of such a media environment, shaped by a wave of media consolidation, can be seen in the number of companies controlling US media dropping from 'fifty to between five and seven in just ten years'. The impact has been in 'the merger of news biz and show biz'. Entertainment-oriented shows help to 'depoliticise viewers, while sensation-driven cable news limits analytical journalism and in-depth issue-oriented coverage.'[43]

One of the major motives for such a development was, as indicated by Kumar, the aftermath of the Vietnam War, when a section of political elites came to believe that it was media coverage of the war that led to the US defeat. Among other things, they argued that television distorted the war by showing graphic images of the dead, turning Americans against the war.[44] Therefore, according to Kumar, the convergence of media and government interests in war propaganda derives from shared economic and political interests, which consequently results in further integration of the media into the military industrial complex, building upon the existing Cold War relationship. This integration and the pressure to increase profit by giant media conglomerates has led to methods of operation that have compromised journalists' ethics. 'The Fox effect shows how this works.'[45]

Media strategies: the embedding policy

According to Christiane Eilders, the method of embedding journalists in military units was one of the US government's new strategies to control information. This strategy had replaced the pool system, adopted since the Vietnam War but used mainly during the 1991 Gulf War, in which journalists had been grouped and led by the military, which consequently affected the media coverage through censorship and restricted movements of journalists in the actual fighting.[46]

Schechter assumed that the embedding experience grew out of the experiences during the Afghan War, where there were clashes between journalists who wanted access to the story on the ground and the military units who physically threatened media representatives. That was stage-managed from the central command's military base in Tampa, Florida, which had been deliberately moved out of Washington to ensure secrecy.[47] Therefore, to avoid extra pressure from the media, the Pentagon offered the media the chance during the war on Iraq to embed their reporters in designated military units, though only some of these were in the main invasion force.

This method also included the journalists undergoing a mild version of basic training to prepare them for battlefield discipline.

As stated by Rutherford, embedded journalists also had to sign contracts to signify their willingness to self-censor, so that information deemed vital to the ongoing military operations would not be released in their reports. Close-ups of dead or wounded soldiers, for example, were taboo, at least until next of kin were notified. In return, the Pentagon and the military would do their best to make sure these embedded reporters could send their stories and images back to their news outlets.[48]

According to Kumar, embedded reporters were telling the story both physically and ideologically from the vantage point of the US and British troops; 'after all, they ate with them, they slept together, and they even wore the same clothes.'[49]

BBC reporter Ben Brown pointed out this identification as he stated in relation to his experience:

> There was an Iraqi who ... jumped up with an RPG and he was about to fire it at us because we were just standing there and this other [warrior] just shot him with their big machine gun and there was a big hole in his chest. That was the closest I felt to being almost too close to the troops ... because if he had not been there he would have killed us and ... afterwards I sought out the gunner who had done that and shook his hand.[50]

Accordingly, this had led to the militarisation of the media, as suggested by Walid Shmait who argued that such a concept was generated by the amount of reports coming from the battlefields focusing on the vocabulary and the technicality of the military, rather than critically questioning the events and the incidents. This, according to Shmait, dominated the coverage of the war and provided a limited view of the war.[51]

Selling the war and the use of language

In its strategy to win international support for invading Iraq, the US administration implemented extensive diplomatic efforts to address the UN Security Council and to assemble convincing evidence about the dangers posed by the Iraqi regime, in order to build an international coalition. The significance of this strategy was to portray the war on Iraq as not only an American war, and to convince people that it was the will of the international community to remove Saddam Hussein. This was encapsulated in a phrase used by President Bush in naming his administration's strategy to win international support – 'the coalition of the willing.'[52]

According to Steve Schifferes of the BBC, on 18 March 2003 the US named thirty countries that had decided to associate their efforts with the US action in Iraq, though few of these countries were providing a major military presence in the Gulf, notably Britain and Australia. In addition there were fifteen countries providing assistance, such as over-flight rights, but which did not want to declare their support.[53]

Nevertheless, Laura McClure noted that in order to build this collaboration, the US administration used a strategy of offering large amounts of foreign aid in exchange for supporting the Iraq War: this strategy was described by McClure as 'the US brandishing its wallet as a weapon'. Turkey, for instance, was offered $6 billion in direct aid, plus billions more in loans, if it would allow the US to base soldiers there in advance of the invasion, and other nations like Guinea, Mexico, Chile, Angola, Cameroon and Pakistan – which were the six undecided countries of the fifteen members of the UN Security Council – faced the dilemma of whether to ignore the 'mounting opposition to war at home, or face the wrath of Washington'.[54]

However, according to Andrew Calabrese the principal arguments offered to the UN Security Council for why the US and UK should invade Iraq were twofold:

> One being that the regime of Saddam Hussein had continued to store, produce, and find ways to further develop the capacity to produce biological, chemical, and nuclear 'weapons of mass destruction' (WMDs) and the other being that there were covert links between the Iraqi government and members of the Al-Qa'ida network, perhaps even implicating Iraqi in the terrorist attack on US targets on September 11, 2001.[55]

As stated by Calabrese, US Secretary of State Colin Powell appeared before the UN Security Council on 5 February 2003 and presented what he characterised as 'compelling evidence of the existence of WMDs in Iraq and of links between al-Qa'ida and Saddam Hussein's regime'. Powell stated that Saddam Hussein had the ability to deliver lethal poisons and diseases in ways that can cause massive [levels of] death and destruction.'[56] The same language was used one day later when President Bush addressed the American nation, stating that the Iraqi regime had 'acquired and tested the means to deliver weapons of mass destruction', including spray devices on 'unnamed aerial vehicles' which, if launched from a ship off the American coast, 'could reach hundreds of miles inland'. Also, Bush made sure on the same occasion to link Saddam Hussein's regime to al-Qa'ida, stating that there are 'longstanding, direct and continuing ties to terrorist networks … [Therefore] the danger Saddam Hussein poses reaches across the world.'[57] However, as indicated by Calabrese, such claims were disputed, and the evidence used to support them was discredited 'before, during and since the US invasion of Iraq'.[58]

Similarly, Britain's Tony Blair, according to David Miller, was very careful in his use of language that 'exploited the media thirst for dramatic threats.'[59] In a key address to the House of Commons Liaison Committee, Blair stated, 'I think it is important that we do everything we can to try to show people the link between the issue of weapons of mass destruction and these international terrorist groups, mainly linked to al-Qa'ida.'[60]

Nevertheless, Miller considered that the attack on Iraq showed 'the integration of propaganda and lying into the core of government strategy', which was coordinated by 'Downing Street, the White House and the Pentagon' in order to make invading and occupying a sovereign country a successful operation.[61] In agreement with Miller, other scholars have also pointed out the use of language, and criticised the way the US administration used language to promote its policies towards war on Iraq. Douglas Kellner, for instance, accused the administration of constantly lying about Iraq and other political issues. Kellner stated that the Bush administration has practised 'the Goebbels-Hitler strategy of the Big Lie, assuming that if you repeat a slogan or idea enough times, the public would come to believe it – that the words would become reality.'[62] Therefore, invading Iraq involved the continual repetition of simplistic slogans aimed to mobilise conservative support without regard to truth.[63] Similarly, Rampton and Stauber pointed out the use of the phrase 'Axis of Evil' by Bush on 29 January 2002, in which he included Iraq, Iran and North Korea, whom he accused of destabilising world peace. The use of the term suggested an alliance or confederation of states that pose a significant danger precisely because of their common alignment, which in reality is not true, as Iran and Iraq had been bitter adversaries for decades and there was no pattern of collaboration between North Korea and the other two states.[64] Likewise, Rampton and Stauber illustrated other terms and phrases used by the US administration such as 'Operation Iraqi Freedom' – the American military campaign against Iraq, which served as a powerful framing device, particularly when television networks, including Fox and MSNBC, used it as their tag line for the war.[65] Other phrases, like 'old Europe', were also assumed to be propagandistic since it was aiming to marginalise the role of Germany and France in opposing the war on Iraq. The term was used by Secretary of Defence Donald Rumsfeld on 22 January 2003 in dismissing French and German insistence that 'everything must be done to avoid war' with Iraq, saying that most European countries stood with the US in its campaign against the Iraqi regime.

> Germany and France represent old Europe, and the NATO expansion in recent years means the centre of gravity is shifting to the East.[66]

It is believed that the use of such language during the war extended beyond selling the concept through focusing on the brutality of Saddam's regime and his military arsenal or links to terrorist groups, to frame certain incidents with phrases that would not reveal the awfulness of the war's reality, and at the same time would appeal to the public as a positive approach, as in such phrases as 'precision bombing', 'surgical strikes' and 'friendly fire'. Sainath argued that though some of these phrases had been used in previous conflicts, they were still common during the run-up to the Iraq war and served alongside other propaganda techniques to shape public opinion.[67]

According to Nancy Snow, the information about such a war will continue to be defined by the control of language from the top. In the case of Iraq, slogans and facile statements about freedom over tyranny from President Bush seemed to satisfy the appetite of the press, while opposing thought from the grassroots required evidence beyond reasonable doubt.[68] (Snow 2003: 81).

Nevertheless, Rutherford stated that control of language of the war was one of the first priorities of the propaganda campaign launched to sell the policy of going to war. According to him, the Pentagon's struggle to get 'the right words' such as 'collateral damage' (meaning civilian deaths) and 'friendly fire' (killing your own) in earlier wars was in itself not new. However, the new twist this time was the effort amounted to particular applications of a new technique called viral marketing that had been born out of the experience of selling on the Internet.

'The idea was to get any promotion to propagate itself by infecting other messages, much as a virus can spread rapidly through an unsuspecting population.'[69] To reach this 'unsuspecting population', Rutherford assumed that the White House and Pentagon were keen to get 'reporters and thus consumers to utilise the right terminology' in order to ensure the right spin on events.[70]

Thus, Rutherford believes that the Pentagon has left its mark on the vocabulary of journalism. He points out that the embedded reporters even produced a form of 'embedded language', which Geoffrey Nunberg of the *New York Times*, described as 'the metallic clatter of modern military lingo ... such as "asymmetric warfare", "emerging targets", and "catastrophic success"'.[71]

Overloading the media and source attribution

According to Schechter, information operations (IO) works in conflicts by providing too much information to the press, which he suggests was the key information mechanism of control during the Kosovo War.[72] Similarly, during the 2003 Iraq War, one of the greatest problems for broadcasters was, according to a 2004 Cardiff School of Journalism report, 'the chaotic volume of information' provided by the embedded journalists.[73] This was illustrated in the way that it was more difficult for broadcasters to deal with than the controlled outflow from the military briefings. 'In many cases, the quality of information coming from the central command was less reliable than information coming from reporters on the ground.'[74] These observations were made from reporting of the battles of Umm Qasr and Nasiriyah and reporting of the checkpoints shootings near al-Najaf.

However, the issue of overloading the media led in certain incidences to further confusion, when attribution was most likely to get lost in the abbreviated language of the headlines and summaries. With reference to the Cardiff School of Journalism study, the issue was due to 'too much faith' in claims coming from the military sources.[75] This faith in such information had lead to additional chaos in reporting the war or, as Kari Lydersen pointed out, 'truth' became one of the major casualties of the media's unquestioning reliance on government sources.[76]

According to Kumar, a large part of psychological operations was the spread of misinformation; this strategy took into consideration the twenty-four-hour news channels' constant demand for new information. Therefore the credibility associated with official sources meant that military claims would often be relayed with no one taking the time to check the facts. 'An update from a military official would receive wide publicity, only to be retracted or modified.'[77] This was observed in many incidents including the Basra uprising, which was labelled by a British forces spokesman, Group Captain Al Lockwood, a 'popular uprising', which was denied by the Iraqi forces. Similarly, the southern Iraqi port of Umm Qasr was reported as being taken 'nine times', while reports on the discovery of chemical weapons at al-Najaf were never confirmed.[78] Furthermore, as cited by Ciar Byrne of the *Guardian*, Richard Sambrook admitted that 'it was proving difficult for journalists in Iraq to distinguish truth from false reports, and that the pressure facing reporters on twenty-four-hour news channels had led to premature [and] inaccurate stories.'[79] Nevertheless, one senior BBC news source commented, 'we are absolutely sick and tired of putting things out and finding out they are not true. The misinformation in this war is far away worse than any conflict I have covered, including the first Gulf War and Kosovo.'[80]

PR involvement: establishing a scene creates a story

As observed by Laura Miller, John Stauber and Sheldon Rampton, the techniques used to sell the war in Iraq were familiar PR strategies.[81] In addition, relevant information had been flown to the media and the public through a limited number of 'well-trained messengers', including seemingly independent third parties.[82] Symptomatic of this, the *Washington Post* reported in July 2002 that the White House had created its Office for Global Communications (OGC), to 'coordinate the administration's foreign policy message and supervise America's image abroad.'[83] On the same subject, in September 2002 *The Times* of London reported that the OGC would spend $200 million for a 'PR blitz against Saddam Hussein' aimed 'at American and foreign audiences, particularly in Arab nations sceptical of the US policy in the region'. The campaign would use 'techniques to persuade crucial target groups that the Iraqi leader must be ousted'.[84] According to Miller et al, other efforts contributed to advance the campaign, such as the media training of Iraqi dissidents by the state department to 'help make the Bush administration's argument for the removal of Saddam'[85] and the use of an informal 'strategic communications' group of 'Beltway lobbyists, PR people and Republican insiders by Secretary of Defence Donald Rumsfeld to sharpen the Pentagon's message'.[86]

Hiring PR groups seemed to be a key element in shaping the whole campaign, especially if these groups were hired on a non-bid basis which, as stated by Miller et al, revealed the government's determination to connect with firms that were 'already versed in running oversees propaganda operations'.[87] These firms were such as the Rendon Group, which, according to its website, has worked in a number of countries, providing discreet and confidential strategic guidance to clients in Africa, the Americas, Asia, Europe and the Middle East. Their clients are 'government agencies, private sector enterprises and non-governmental organisations that face the challenge of achieving information superiority in order to impact on public opinion and outcomes.'[88]

Accordingly, Rampton and Stauber suggested in their article 'How to sell a war' in *In these Times* magazine, that some images of the war on Iraq may have been cooked up by PR specialists and 'perception managers'.[89] As stated by Rampton and Stauber, 'visual images are what most people would remember of conflicts.'[90]

Interestingly, Cardiff School of Journalism's study on coverage of the Iraq War found that the most remembered images of the conflict were the toppling of Saddam's statue at al-Ferdaws Square, and the rescue of Private Jessica Lynch.[91]

The study also indicated that the former incident was widely referred to in focus groups, although reactions to it were divided. Many saw it as an accurate reflection of Iraqi joy at being liberated, while others felt there was a degree of overacting for the cameras. There was also some concern at the tightness of the camera's focus on the crowd, and whether this painted an accurate picture of Baghdad, and of Iraq as a whole.[92] This is perhaps what led Rampton and Stauber to pose the question of whether the toppling of Saddam's statue was as spontaneous as it was made to appear, or if there was a reason the scene was 'a bit too picture-perfect', hinting at the role of the PR machine in constructing it.[93]

In regard to Private Jessica Lynch, the Cardiff study indicated that the second most recalled incident was her 'rescue', which was remembered by 44 per cent of respondents to the question. This too, among focus groups, was discussed with a healthy dose of scepticism. Strong statements of distaste and disbelief were expressed. 'More significantly, this incident was linked to the "unreality" of the action images, with the comparison with Hollywood movies again being drawn.'[94]

On the other hand, in an interview with this author, Ibrahim Helal, head of news at al-Jazeera during the war on Iraq, expressed his scepticism of the images of the toppling of the statue of Saddam, while he stated that the images of Jessica's rescue seemed more convincing.

> I remember I was in the newsroom during the toppling of Saddam's statue, and we felt it was staged, especially when an American soldier took the American flag from his pocket to cover the face of the statue. The Jessica story was more convincing; it could happen during the course of any war. We did not have time to check the credibility of the story – remember, we were functioning around the clock, so in this matter we had to cope with the speed of the incidents as they came in, otherwise we would be outdated.[95]

This fast flow of incidents during the war is normally one of the key elements that many media strategists take into consideration in the twenty-four-hour news environment. It would normally put huge pressure on the media in order to keep up with the sequence of events without losing its presence in the field. During the war on Iraq, many 'facts' and pieces of evidence were stated as truth, only to be recanted later when their veracity was called into question.[96]

Conclusion

Some of the most significant aspects of the war were the particular ways the media was integrated into the war effort, as never before seen in conflict.

The use of sophisticated technology during this war, and hence the implementation of advanced techniques in communication strategy, elevated the discussion on the role of media in modern conflicts. This strategy included the embedding policy aimed at ensuring favourable reporting of the military. It is clear that the military understood the lessons of previous conflicts and had therefore developed a more sophisticated system to control information, promoting self-censorship among embedded journalists through the psychological effects caused by bonding with the units they were travelling with, whilst eating, sleeping and working with them.

Also remarkable is the role of media moguls, who contributed to the official media strategies drawn up by the White House and the Pentagon. I would argue that their common interests shaped this relationship, which was reflected in the growth of certain media outlets.

The remarkable involvement of the PR industry in the war contributed in some cases to the misinformation techniques adopted by the White House and the Pentagon for military and security reasons. Stories like the saving of Private Jessica Lynch and the toppling of the Saddam statue proved to be orchestrated by PR and public affairs officers. Equally important was the use of language and terminology, which was carefully used in addressing the public and the media. Emotive phrases like 'coalition of the willing', 'friendly fire' 'old Europe' and 'surgical strikes' were all echoed by the media and served to shape a supportive public opinion.

Techniques and methods from the media strategies implemented in the Iraq War and accompanying information technology will continue to play a prime role in wartime propaganda, and they will become even more sophisticated, employing new technologies as they advance. Media professionals will face greater challenges with the concentration of ownership of media organisations, as commercial and economic agendas become intertwined with political aims to control the public perception of conflicts. It is, however, the journalists' experiences and awareness of future challenges which could create a window for professional progress in serving the public's right to know other aspects of war and conflict.

Notes

1. Bell, M., *Through Gates of Fire* (TGF), p. 193.
2. Brown, R., 'Spinning the War: Political Communications, Information Operations and Public Diplomacy in the War on Terrorism', p. 90.
3. Ibid.
4. Ibid, p. 91.
5. Ibid.
6. Lewis, J., Brooks, R., Mosdell, N. and Threadgold, T., *Shoot First, Ask the Question Later: Media Coverage of the 2003 Iraq War*, pp. 16-17.
7. Miller, D., 'Information Dominance: the Philosophy of Total Propaganda' (IDPTP), in *War, Media, and Propaganda*, p. 7.
8. Ibid, p. 8.
9. Ibid.
10. Ibid, p. 10.
11. Ibid.
12. TGF, pp. 198-200.
13. Katovsky, B. and Carlson, T., *Embedded: The Media at War in Iraq*, p. xi.
14. Tumber, H. and Webster, F., *Journalists Under Fire: Information War and Journalistic Practices*, p. 21.
15. Seib, P., *Beyond the Front Lines: How the News Media Cover a World Shaped by War*, p. 53.
16. Ibid.
17. The first embed journalist was *The Washington Post*'s columnist and *Atlantic*'s monthly editor and columnist Michel Kelly, who died in a Humvee crash while riding with the US Army's 3rd Infantry Division at south Baghdad airport. The second was the NBC news correspondent David Bloom who died of pulmonary embolism while travelling with the US Army's 3rd Infantry Division outside Baghdad (www.freemedia.at). Katovsky, B, 2003, p. xiv.
18. http://www.freemedia.at
19. RSF 2006: 2.
20. Ibid. The Paris-based Reporters without Borders (RSF) said in October 2007 that fifty-two journalists and media workers have been killed in Iraq since the beginning of 2007, while the International Federation of journalists (IFJ) said that a total of 234 journalists have been killed in the country since the US-led invasion in March 2003 (www.iraqupdates.com).
21. According to the International Press Institute Report for 2003 death watch, 'Tariq Ayoub, a cameraman and correspondent for the Qatar-based satellite television network, Al-Jazeera, was killed during a US air raid on Baghdad. Ayoub, a Jordanian citizen, died in hospital after he was wounded in the strike, which set ablaze Al-Jazeera's office near the Information Ministry,

the network said. Another member of Al-Jazeera's Baghdad crew, Zohair al-Iraqi, was wounded in the attack.' (http://www.freemedia.at).

The report also stated that on the same day (8 April 2003) 'Taras Protsyuk, a Reuters cameraman, and Jose Couso, a cameraman for the Spanish television channel Tele 5 were killed when a US tank fired a shell at the Palestine Hotel, the base for many foreign media in Baghdad. Protsyuk, a Ukrainian based in Warsaw, was killed immediately; Couso was wounded and later died in hospital. Three other members of the Reuters team in Baghdad were hurt in the tank shell blast.' (ibid).

22. Simpson, J., *The Wars Against Saddam: Taking the Hard Road to Baghdad*, p. 345.
23. Ibid, p. 315.
24. Lewis, J., Brooks, R., Mosdell, N. and Threadgold, T., *Shoot First, Ask the Question Later: Media Coverage of the 2003 Iraq War*, pp. 22-23.
25. Seib, P., *Beyond the Front Lines: How the News Media Cover a World Shaped by War*, p. 125.
26. Tumber, H. and Palmer, J., *Media at War: The Iraq Crisis*, p. 64.
27. Ibid, p. 64.
28. Ibid.
29. Taverner cited, ibid, pp. 64-65.
30. Tatham, S., *Losing Arab Hearts and Minds: The Coalition, Al-Jazeera and Muslim Public Opinion*, p. 86.
31. Ibid, p. 88.
32. Rutherford, P., *Weapons of Mass Persuasion*, pp. 61-62.
33. Ibid.
34. Gutierrez, Miren, 'The 'Prop-Agenda' at War', np.
35. Ibid.
36. Schechter, D., 'Selling the war: The media management strategies we never saw', p. 26.
37. Ibid, pp. 25-26.
38. Kumar, D., 'Media, War, and Propaganda: strategies of information management during the 2003 Iraq War', p. 49. Furthermore, Kumar stated that several new facets of censorship were incorporated, making the system more sophisticated.

In part, this evolution and perfecting of the media-military industrial complex propaganda system is voluntary and conscious. For instance, Bush advisors Karl Rove and Mark McKinnon met with the heads of Viacom, Disney, MGM and others after 9/11 to discus how the media could 'help' the government's efforts. Before the start of the Iraq War, CNN set up a system of 'script approval' where reporters had to send their stories to unnamed officials in Atlanta before they could be run. This would ensure if the military made any errors, CNN monitors would act as the second layer of filtering. Rupert Murdoch of News Corporation took an active role in setting the tone of his news media outlet, so that, not coincidentally, all 175 editors of Murdoch's worldwide newspaper empire took a position in support of the war. Fox, also owned by News Corporation, took this support to the extreme, going so far as to ridicule antiwar protesters. Kumar, D., 'Media, War, and Propaganda: strategies of information management during the 2003 Iraq War', p. 51.

39. Ibid, p. 49.
40. Solomon, N., 'The Military-Industrial-Media Complex: Why war is covered from the warriors' perspectives', np.
41. Ibid.
42. Ibid.
43. Schechter, D., 'The Media, the War and Our Right to Know', np.
44. Kumar, D., 'Media, War, and Propaganda: strategies of information management during the 2003 Iraq War', p. 60.
45. Ibid, p. 52.
46. Eilders, C., 'Media under fire: Fact and fiction in conditions of war', pp. 643–4.
47. Schechter, D., 'Selling the war: The media management strategies we never saw', p. 26.
48. Rutherford, P., *Weapons of Mass Persuasion*, p. 72.
49. Kumar, D., 'Media, War, and Propaganda: strategies of information management during the 2003 Iraq War', p. 61.
50. Miller, D., 'Information dominance: The Philosophy of total propaganda', in *War, Media, and Propaganda*, p. 10.
51. Shmait, W., *Imbratoreyat Al-Mohafizeen Al Jodod: Al Tadleel Al-Ilami wa Harb Al-Iraq* (The Empire of Neoconservatives: Media Deception and the Iraq War), pp. 227-228.
52. The term 'coalition of the willing' is a post-1990 political phrase used to describe military or military/humanitarian interventions for which the United Nations Security Council cannot agree to mount full UN peacekeeping operations. It has existed in the political science/international relations literature since UN peacekeeping operations began to run into deep trouble in 1993–4, and alternatives began to be considered. It has been applied to the NATO intervention in Kosovo ('illegal but legitimate'), the Australian-led INTERFET operation in East Timor, and its most well-known example, the US-led invasion of Iraq in March 2003 (wikipedia.org).
53. Schifferes, S., 'US names "coalition of the willing"', *BBC News Online*, 18 March 2003, np.
54. McClure, L., 'Coalition of the billing – or unwilling?', *Salon.com*, 12 March 2003, np. As stated by McClure, the Bush administration required nine

votes from among the fifteen members of the UN Security Council to authorise the invasion, therefore its policy was to influence the undecided six countries' decision by offering them economic and financial support.

55. Calabrese, A., 'US media and the justification of the war on Iraq, Television and New Media', 2005, no. 6, May, issue 2, p. 155.
56. Ibid, p. 156.
57. Ibid, p. 165.
58. Ibid.
59. Miller, D., 'Introduction', in *Tell me lies: Propaganda and Media Distortion in the Attack on Iraq,* Pluto Press, London, 2004b, p. 5.
60. Ibid.
61. Ibid.
62. Kellner, D., 'Lying in Politics: the case of George W. Bush and Iraq', *Cultural Studies-Critical Methodologies,* 2007, no.7, issue 2, p. 135.
63. Ibid.
64. Rampton, S. and Stauber, J., 'How to sell a War: The Rendon Group deploys "perception management" in the war on Iraq', *In These Times,* 8 April 2003, p. 115.
65. Ibid.
66. *CNN Online,* 'Rumsfeld: France, Germany are 'problems' in Iraq conflict', 23 January 2003, np.
67. Sainath, P., 'War, media, propaganda and language: Coalition of the killing', *The Hindu,* 13 April 2003, np.
68. Snow, N., *Information War: American Propaganda and Opinion Control Since 9–11,* Seven Stories Press, New York, 2003, p. 81.
69. Rutherford, P., *Weapons of Mass Persuasion,* p. 62.
70. Ibid.
71. Ibid, p. 64.
72. Schechter, D., 'The Media, the war and Our Right to Know', *Alternet.org,* 1 May 2003, p. 30.
73. Lewis, J., Brooks, R., Mosdell, N. and Threadgold, T., *Shoot First, Ask the Question Later: Media Coverage of the 2003 Iraq War,* Peter Lang, New York, 2006, p. 29.
74. Ibid.
75. Ibid.
76. Lydersen, K., 'An Army of Propaganda', *Alternet.org,* 31.03.2003, np. In her article, Lydersen illustrates two such incidents; one was on 20 March, when reporters from NBC, NPR, ABC and other outlets reported as fact the military's assertion that the Iraqis had used banned Scud Missiles. 'However, two days later the Joint Chiefs of Staff reported that in fact no Scud Missiles had been fired.' The other was on 23 March when various media trumpeted the government's claim that a chemical weapons factory

had been found near the town of Najaf, 'though a day later that claim was totally debunked'.
77. Kumar, D., 'Media, War, and Propaganda: strategies of information management during the 2003 Iraq War', p. 62.
78. Byrne, C., 'BBC chiefs stress need to attribute war sources', *Guardian*, 28 March 2003, np.
79. Ibid.
80. Ibid.
81. Miller, L., Stauber, and J., Rampton, S., 'War is Sell', in *Tell me lies: Propaganda and Media Distortion in the Attack on Iraq*, Pluto Press, London, 2004, p. 44.
82. Ibid.
83. Ibid, p. 45.
84. Ibid.
85. Ibid, p. 46.
86. Ibid.
87. Ibid, p. 42.
88. Rendongroup.com, 2006: np. As it said on its website 'To date, we have worked in 91 countries planning and managing strategic and tactical communications programs across Africa, the Americas, Asia, Europe, and the Middle East. The global experience of our seasoned staff maximizes the precision of our analysis and the value of our counsel. The Rendon Group office in Washington, DC operates a 24/7 media monitoring watch-centre in support of both clients and TRG personnel working on location worldwide.'
89. Rampton, S. and Stauber, J., 'How to sell a War: The Rendon Group deploys "perception management" in the war on Iraq', *In These Times*, 8 April 2003, np.
90. Ibid.
91. Lewis, J., Brooks, R., Mosdell, N. and Threadgold, T., *Shoot First Ask the Question Later: Media Coverage of the 2003 Iraq War*, p. 38. 'When asked an open-ended question about what they remembered from the coverage, people responded with a wide variety of recollections. Top of the list was the toppling of the statue of Saddam Hussein as US troops took Baghdad. In response to subsequent, more specific questions, 80 per cent said they remembered this incident very well (and only 3 per cent not at all), confirming the impact made by this moment. While it could be argued that this is simply a memorable moment, it seems likely that this high level of recall also reflects the volume of media coverage afforded to it.'
92. Ibid.
93. Rampton, S. and Stauber, J., 'How to sell a War: The Rendon Group

deploys "perception management" in the war on Iraq', *In These Times*, 8 April 2003, np.
94. Lewis, J., Brooks, R., Mosdell, N. and Threadgold, T., *Shoot First, Ask the Question Later: Media Coverage of the 2003 Iraq War*, pp. 38-39.
95. Helal, I., Head of News at Al-Jazeera during Iraq war 2003, interviewed on 15 April 2004 on the phone.
96. Kumar, D., 'Media, War, and Propaganda: strategies of information management during the 2003 Iraq War', p. 52.

Bibliography

Bell, M., *Through Gates of Fire: A Journey Into World Disorder*, Weidenfeld and Nicolson, London, 2003.

Brown, R., 'Spinning the War: Political Communications, Information Operations and Public Diplomacy in the War on Terrorism', in *War and the Media*, Sage, London, 2003.

Katovsky, B. and Carlson, T., *Embedded: The Media at War in Iraq*, The Lyons Press, Guilford, CT, 2003.

Lewis, J., Brooks, R., Mosdell, N. and Threadgold, T., *Shoot First, Ask the Question Later: Media Coverage of the 2003 Iraq War*, Peter Lang, New York, 2006.

Miller, D., 'Information Dominance: the Philosophy of Total Propaganda', in *War, Media, and Propaganda*, Rowman and Littlefield, Lanham, Maryland, 2004a.

Miller, D., 'Introduction', in *Tell Me Lies: Propaganda and Media Distortion in the Attack on Iraq*, Pluto Press, London, 2004b.

Miller, L., Stauber, J., Rampton, S., 'War is Sell', in *Tell Me Lies: Propaganda and Media Distortion in the Attack on Iraq*, Pluto Press, London, 2004.

Rampton, S. and Stauber, J., *Weapons of Mass Deception: The Use of Propaganda in Bush's War on Iraq*, Constable and Robinson, London, 2003.

Rutherford, P., *Weapons of Mass Persuasion*, University of Toronto Press, Toronto, 2004.

Schechter, D., 'Selling the War: the Media Management Strategies We Never Saw', in *War, Media, and Propaganda*, Rowman and Littlefield, Lanham, Maryland, 2004.

Seib, P., *Beyond the Front Lines: How the News Media Cover a World Shaped by War*, Palgrave Macmillan, New York, 2004.

Shmait, W., *Imbratoreyat Al-Mohafizeen Al Jodod: Al Tadleel Al-Ilami wa Harb Al-Iraq* (The Empire of Neoconservatives: Media Deception and the Iraq War), Al-Saqi, London/ Beirut, 2005.

Simpson, J., *The Wars Against Saddam: Taking the Hard Road to Baghdad*, Macmillan, London, 2003.

Snow, N., *Information War: American Propaganda and Opinion Control Since 9–11*, Seven Stories Press, New York, 2003.
Tatham, S., *Losing Arab Hearts and Minds: The Coalition, Al-Jazeera and Muslim Public Opinion*, Hurst, London, 2006.
Tumber, H. and Palmer, J., *Media at War: The Iraq Crisis*, Sage, London, 2004.
Tumber, H. and Webster, F., *Journalists Under Fire: Information War and Journalistic Practices*, Sage, London, 2006.

Journals and Reports:
Calabrese, A., 'US media and the justification of the war on Iraq, Television and New Media', 2005, no. 6, May, issue 2, pp. 153–175.
Eilders, C., 'Media under fire: Fact and fiction in conditions of war', *International Review of the Red Cross*, 2005, no. 87, December, issue 860, pp. 639–648.
Kellner, D., 'Lying in Politics: the case of George W. Bush and Iraq', *Cultural Studies-Critical Methodologies*, 2007, no.7, issue 2, pp. 132–144.
Kumar, D., 'Media, War, and Propaganda: Strategies of Information Management during the 2003 Iraq War', *Communication and Critical/Cultural Studies*, 2006, no. 3, March, issue 1, pp. 48–69.
Lewis, J., et al, 'Too close to comfort? The role of embedded reporting during the 2003 Iraq war: summary report', *Cardiff School of Journalism Media and Cultural Studies*, 2004.
International Press Institute, '2003 Death Watch', 2003, *I.P.I.* [http://www.freemedia.at/cms/ipi/deathwatch.html?year=2003 (Accessed 1 November 2007)].
Reporters Without Borders, 'Slaughter in Iraq', March 2006, *RSF*, [http://www.rsf.org/article.php3?id_article=16793 (Accessed 1 February 2007)].

Online Sources:
Byrne, C., 'BBC chiefs stress need to attribute war sources', *Guardian*, 28 March 2003, [http://www.guardian.co.uk/media/2003/mar/28/broadcasting.Iraqandthemedia1 (Accessed 17 February 2008)].
CNN Online, 'Rumsfeld: France, Germany are 'problems' in Iraq conflict', 23 January 2003, [http://www.cnn.com/2003/WORLD/meast/01/22/sprj.irq.warp (Accessed 17 February 2008)].
Gutierrez, Miren, 'The 'Prop-Agenda' at War', *Inter Press Service News Agency (IPS)*, 27 June 2004, [http://www.ipsnews.net/interna.asp?idnews=24386 (Accessed 6 January 2007)].
Lydersen, K., 'An Army of Propaganda', *Alternet.org*, 31 March 2003, [http://www.alternet.org/story/15507 (Accessed 20 November 2006)].
McClure, L., 'Coalition of the billing – or unwilling?', *Salon.com*, 12.03.2003, [http://www.dir.salon.com/story/news/feature/2003/03/12/foregin_aid/index.html (Accessed 15 February 2007)].
Rampton, S. and Stauber, J., 'How to sell a War: The Rendon Group deploys

"perception management" in the war on Iraq', *In These Times*, 8 April 2003, [http://www.inthesetimes.com/comments.php?id=299_0_1_0_C (Accessed 6 January 2007)].

Sainath, P., 'War, media, propaganda and language: Coalition of the killing', *The Hindu*, 13 April 2003, [http://www.thehendu.com/thehindu/mag.2003/04/13/stories/2003041300100100.htm (Accessed 17 February 2008)].

Schechter, D., 'The Media, the War and Our Right to Know', *Alternet.org*, 1 May 2003, [http://www.alternet.org/story/15797 (Accessed 2 November 2007)].

Schifferes, S., 'US names 'coalition of the willing'', *BBC News Online*, 18 March 2003, [http://www.news.bbc.co.uk/2/hi/americas/2862343.stm (Accessed 14 February 2007)].

Solomon, N., 'The Military-Industrial-Media Complex: Why war is covered from the warriors' perspectives', *Fair.org*, July/August 2005, [http://www.fair.org/index.php?page=2627 (Accessed 7 February 2007)].

The Rendon Group, [http://www.rendon.com (Accessed 15 February 2008)].

Wikipedia online, [http://www.wikipeida.org (Accessed 14 February 2008)].

Interview:

Helal, I., Head of News at Al-Jazeera during Iraq war 2003, interviewed on 15 April 2004 on the phone.

OLFA LAMLOUM

Al-Jazeera: a Pan-Arab Revival?

Published in August 2007, the Annual Arab Public Opinion Survey confirmed for the third consecutive year two facts that seemed from then on established in the Arab world. The first concerned strong anti-American sentiments; the second was the primacy of the satellite network al-Jazeera as the most important source of information (54 per cent in 2006), by far exceeding competition from the Saudi station al-Arabiya (5 per cent), and the American, al-Hurra (1 per cent).[1] It goes without saying that this type of survey hardly provides a grasp of the complexity of the social dynamics underlying these two facts, and even less so their evolution and impact. It serves, on the other hand, to remind us that the Qatari news channel continues to produce an endogenous journalistic narrative with a strong appeal that developed in the course of crises and conflicts in which the United States was fully or partially engaged.

October–November 2001 in Afghanistan, March 2003 in Iraq, summer 2006 in Lebanon, and March 2008 in Gaza; the conflicts succeed each other while the extra-territoriality of the new information and communication technologies overturns the conditions of production, circulation and reception of news in the Arab world. In an increasingly competitive media environment, al-Jazeera nevertheless succeeds in distinguishing itself. Its coverage, exclusive (in the case of Afghanistan) or highly sustained (in the case of Iraq, Lebanon and Gaza),[2] allowed it to break the monopoly of the western media. These 'moments of mobilisation'[3] allowed the station to impose itself as a source of news and images. It thereby earned status as a producer of new transnational flows of communication, competitive in terms of the construction of 'the truth' and the manufacture of 'legitimate' information.

Since its launch in 1996 al-Jazeera has evoked two lines of questioning, the importance of which becomes obvious when we look at its coverage of the official Arab political scene, the second Palestinian intifada and the 11 September attacks in 2001 and their aftermath. The first objective is to understand how a TV station, considered insolent, came to see the light of day in an authoritarian environment. Second comes the question of how the reasons that allowed it to distinguish itself arose; notably, its coverage of the two regional 'seismic epicentres',[4] the Israeli-Palestinian conflict and the invasion of Iraq. Answering these two questions allows us to explore how the Arab power elites reasserted their dominance, and to examine new forms of political mediation and the affirmation of transnational identities.

This article proposes to briefly present the distinctive journalistic narratives of the Qatari station, as they manifested themselves during the US invasion of Iraq in 2003, after which we will discuss the regional political stakes that this distinction hides. Finally, we will broach the question of the Pan-Arabism of which Al-Jazeera is supposed to be the new vector.

The Iraqi ordeal

Without doubt, the war of 2003 offers a very pertinent perspective from which to understand al-Jazeera, to grasp the tensions to which its reports bear witness, and to point out the defining features of the network. Unlike the 2001 war in Afghanistan, the military intervention in Iraq happened on Arab soil, putting the network in close proximity to the conflict, in the context of an exceptional mobilisation of the media. This brings al-Jazeera into fierce competition and exposes it to the strategies of a multitude of actors. The choice of words and images, their articulation, the decoding, the qualification and the layout of the facts were forged on the borders of a number of fields of influence and power.

By the end of 2002, aware of the scope of the impending task, al-Jazeera mobilised exceptional logistical and human resources to cover the war from the principal arenas: Iraq, the US, Great Britain and, finally, the US command centre in Doha. The first place al-Jazeera's apparatus was installed was Iraq. More than thirty journalists and technicians were deployed in Baghdad, Mossul and Busra – where the Qatari station was the only non-Iraqi station on the ground. Al-Jazeera also secured authorisation to embed a journalist with the US troops. Another saw himself allocated to the autonomous Kurdish zone and circulated between Dohouk, Arbil, Souleimaniye and Kirkouk. The second pillar of this operation was set up in the United States, where five correspondents guaranteed continuous coverage from the Pentagon, the White House, the State Department and finally the UN. A team of six journalists covered the American central command centre in Al-Sayliya.[5]

The strong presence of the station in Iraq, moreover benefiting from a certain knowledge of the terrain[6] and from 'favours' granted by Saddam Hussein's regime,[7] allowed an immediacy in its coverage of the conflict which is unequalled in the history of wars in the Arab world. In addition, its diversified geographic deployment offered the station the opportunity to multiply its sources and to extract itself, relatively, from the 'zoom effect' often inherent in media coverage of war zones.

It goes without saying that such an undertaking was only possible thanks to the colossal budget allocated to the coverage of this event. It was facilitated by a Qatari monarchy anxious to guarantee a privileged presence for the network in the Pan-Arab and international media sectors. But if the mastery of digital technology and the scope of the financing were necessary conditions for an existence on the global market of news, it was above all through its individual narrative, contesting the one produced by the dominant western media, that al-Jazeera could distinguish itself. Its position consisted of refuting the rhetoric legitimising the war, denouncing the economic and geo-strategic motives and contesting the strategy of the Bush administration in Iraq and the whole of the Arab world.

On 18 March 2003, two days before the beginning of the military operation, the station expressed its commitment clearly, describing the imminent conflict as 'the war on Iraq'. The unfolding of the war led the station to adjust its language. Starting on 24 March, highlighting its exactitude in the field of semantics, al-Jazeera designated the American and British troops deployed in Iraq 'invading forces' (*al-quwwât al-ghâziya*). It banned from its reports the expression 'coalition forces' used by the army officials and the dominant western media. The Qatari station was not the only media organ opposed to the war – other large European networks and mainly French stations, inspired generally by Jacques Chirac's stance, adopted the same posture. Al-Jazeera, however, went much further than these others by explicitly supporting a defeat of the 'invading forces', a key concept that imposes its own reading of the conflict. This template, which structured and unified its journalistic narrative of the war, progressively imposed itself in the Pan-Arab media, forcing al-Arabiya – launched with mainly Saudi capital in February 2003 with the declared goal to marginalise al-Jazeera – to follow its lead.

In the journalistic narrative of the Qatari network, the qualification of the 'war on Iraq' as a neo-colonial enterprise would become visible through three style elements: the representations of the stakes of the conflict, the treatment of the two principal protagonists and, finally, the way the event was adapted for the screen. At no point during the twenty-one days that the conflict lasted did al-Jazeera use the paradigm of 'democratisation' to report on it. Saddam Hussein kept his title, president, until the fall of Baghdad and he was not referred to as a dictator. Not a single news report, in Baghdad, Mossul or Busra, addressed the question of civil liberties in Iraq. It was never implied that the few civilian testimonies that network correspondents gathered in these three cities required scrutiny. Without doubt this choice could in part be justified by the controls imposed by the Iraqi regime on all foreign media and more drastically on Arabic media. Nevertheless, the explanation lies fundamentally somewhere else. It is found in the editorial choices of the network. The expression *al-ghazu* (the invasion) established the colonial paradigm as the central and quasi-exclusive lens for looking at the events on the Iraqi scene, thereby eclipsing even 'the question of democracy'. Vying to distinguish itself firmly from an agenda that was considered imperial, al-Jazeera voluntarily turned its back on the dominant discourse about the necessary 'democratisation of Iraq' and attempted to produce another interpretation of the conflict. This choice was not without consequences, because in the wake of the military operations, it drew a lot of criticism in Iraq and notably from the Shi'i population.

The second element structuring the narrative is the equal treatment of the information given by the protagonists (the invading forces and the Iraqi authorities). The words of a representative of the American general command were not considered more credible than those of the Iraqi ministry of information. This stance, unique in the field of global media, allowed the network to avoid certain pitfalls of the media war. The most famous episode was the 'insurrection in Busra', which was announced by a British military source. Unlike the BBC and other western TV stations, al-Jazeera not only did not take up the information, but also refuted it.

The third specific element points to the 'script' of the war. Two days after the start of military operations, the theme of 'resistance to the invasion' appeared on the Qatari channel. Until the fall of the capital it served as the main thread of its narrative. Thus, al-Jazeera positioned itself, its images and its recordings[8] not behind the 'coalition forces' that advanced on Baghdad by 'neutralising pockets of resistance' and 'securing zones', but behind the resistance which opposed the invasion. This bias was not without consequence either. It exposed the network to the pressure and the criticism of the Bush administration, but it also allowed for a few slippages in favour of the Iraqi regime.[9]

Even more than during the armed intervention in Afghanistan, al-Jazeera appeared to be subjected to a number of powerful, and at the same time antagonistic, competing and converging strategies. The opposing pressures and strategies of the American administration and the Iraqi and Qatari regimes on one hand, and the journalistic ethics it committed itself to (an opinion and counter-point) on the other, shaped its coverage and demarcated the margins of its autonomy. Nevertheless, throughout this test, al-Jazeera above all saw itself confirmed as principal Pan-Arab publisher of the meaning of the conflict, and as a vector contesting the 'US moment'.[10] The network emerged thus as a social player endowed with the 'power to signify',[11] according to the formula of Stuart Hall.

The twenty-one days of the war also constituted a critical moment during which the American administration changed its assessment of the network. In 1996 the Clinton administration was rather favourable to its launch. Seven years later the Bush administration displayed mounting suspicion towards it, beginning with the interventions in Afghanistan and worsening in the wake of Iraq war.[12] Direct or relayed through the Qatari government, the intimidation and the pressure on the management of the network and on its journalists intensified,[13] to the point at which, in

August 2004, the network was almost banned by the Allawi government as a result of its coverage of the battle of Fallouja, during which it was the only source of information that was not 'filtered by the marines', to use an expression from *The New York Review of Books*.[14]

The persistence of these pressures obliged al-Jazeera to relieve its general director[15] of his duties and to adopt cosmetic measures, destined, notably, to tone down the drama of its appearance and to shorten its talk-show programmes. But the most obvious change concerned the treatment of the situation in Iraq. Al-Jazeera stopped referring to 'occupation forces', instead using 'US armed forces'. Similarly, it treated the post-war period with more distance, giving more time to government activity and, thereafter, to that of groups from the *Sahwa Islamiyya*, the Sunni militias armed by the United States.[16]

In the meantime, the new 'public diplomacy' of the United States, laid out in the National Security Strategy document from 2002,[17] had integrated the new realities of the Arab media landscape. From now on, it attributed major importance to the media effects of its intervention, which translated into the implementation, under the direct control of the US Broadcasting Board of Governors, of a media apparatus directed at the Arab public with the purpose of turning it away from 'hateful' media networks.[18]

Al-Jazeera, or 'elite dissensus and policy uncertainty'

In order to understand the background of its editorial line vis-à-vis the political engagement of the US in the region, one must, without doubt, reassess al-Jazeera in its regional geopolitical context.

As was often highlighted, al-Jazeera is not without historical preecedent. Until the defeat of 1967, *Sawt al-Arab* (the Voice of the Arabs), launched one year after the republican coup d'état of the Egyptian Free Officers Movement in July 1952, was in fact considered a tool for pan-Arab mobilisation against the neo-colonial enterprises in the Arab world. But the comparison ends here because, unlike Nasserism, the al-Thani regime does not make any Arab nationalist claims and al-Jazeera is in no way a reflection of its official ideology. To grasp the recent evolution of the media landscape in the Arab world, Naomi Sakr highlights the pertinence of Daniel C. Hallin's approach in his analysis of the media in the United States during the

Vietnam War. Hallin shows how the political dissent among the power elites in the US and the incertitude hanging over the issue of military intervention in Vietnam allowed for media coverage that spilled over the borders of the 'sphere of legitimate controversy'.[19]

De facto, since the 1990s, the context that guided the proliferation of Arab satellite TV stations reveals the same features.[20] Far from being an act of democratisation from above, the re-composition of the satellite TV sector is above all a demonstration of the tensions and rivalries among the Arab power elites. This process is driven by strategies for regional or local control, more or less competitive and subject to the double pressure of new neoliberal international norms and a strong Islamist opposition. The reasons for the launch of al-Jazeera, following the example of other Saudi, Lebanese or Egyptian networks, do not breach this rule, even if they participate in the construction of a new form of domination.

In some measure, the Qatari station seems to be a sign of the weakening of 'the Saudi era'[21] and reveals the affirmation of new regional players that claim greater visibility and display a new mode of governance. This is also the reason it was perceived as a challenge by the Saudi monarchy. In context of the Cold War, the defeat of Nasserism in 1967 and three years later the death of the *rayis*, rid the monarchy of a strong rival. The crisis of the nationalist model and of the Marxist movements – notably the Palestinian movements in Jordan and the socialist movements in South Yemen and Dhofar (Oman)[22] – allowed Saudi Arabia to impose itself as essential regional player[23]. It was engaged in a number of areas in stemming nationalist and Marxist ideology and instituted itself as kingpin of the regional order inscribed in the Pax Americana. It employed numerous tools and considerable resources. In 1969, it constituted the Organisation of the Islamic Conference (OIC), which played a not negligible role in the promotion of a conservative Islam. After 1979 the monarchy gave its unconditional support to the American-backed mujahedin in Afghanistan in their resistance against the Soviet troops. Similarly it supported Saddam Hussein in his war against the Islamic Republic of Iran. In 1981 the kingdom had drawn its neighbours (Bahrain, the United Arab Emirates, Kuwait, Oman and Qatar) into its orbit of influence by establishing the Gulf Cooperation Council. Saudi accession was equally translated in the media sector using, from the beginning of the 1990, digital technology to progressively spread into the satellite broadcasting sector.[24]

In 1990, in the wake of the Iraqi annexation of Kuwait, Saudi Arabia authorised the return of the US armed forces onto its territory. In this way the kingdom legitimised its implantation in the whole region. Notably, the presence of the Prince Sultan base allowed it to play a major role in the restoration to power of the al-Sabahs in Kuwait, and in the weakening of Iraq.

The liberation of Kuwait from the Iraqi troops and the launch of Israeli-Palestinian negotiations appeared as an uncontested victory of the imperial order. Yet the latter fuelled new resistances. Because, in spite of the direct presence of the US military – a fact that was unprecedented in the region – Washington provoked an increasing hostility from the population. The outgrowth of this hostility in the form of al-Qa'ida and related groups in Iraq, in Lebanon and in the Maghreb, did not spare Saudi Arabia.[25] From the middle of the 1990s the kingdom witnessed on its territory the assertion of a radical Islamist opposition, surprising the world in 2001 when taking into account the number of Saudis participating in the attacks. The emergence of this opposition revealed the weaknesses of the 'Saudi era' and publicly exposed the kingdom to criticism and pressure of the Bush administration.[26]

This new disorder weakened the monarchy on a regional scale, inspired new ambitions of certain neighbours and poisoned the official inter-Arab relations. From the end of the 1990s onwards it inspired a number of Gulf countries to timidly explore new forms of legitimisation and mobilisation. The municipal and legislative elections in Bahrain in 2002, the 'strenuous waking' of the Omani state[27] and finally the local elections in Saudi Arabia in 2005, in spite of their limits, are the proof of the reformative mood among the power elites in these countries. Qatar emerged as a precursor by according women the right to vote, by dissolving its ministry of information, by promulgating employment laws and by adopting a constitution. It is in the context of this evolution that the al-Jazeera network has to be considered. Launched in 1996, it was the product of a specific political moment that did not accord the same role to the Saudi monarchy and allowed new players, escaping Saudi tutelage, to claim new sources of legitimacy. Al-Jazeera is, in that sense, a product of the new Arab disorder. The network insinuated itself into the fissures of this disorder to establish itself as a central element in the Qatari efforts for political reform propagated by Sheikh Hamad Ben Khalifa al-Thani in the wake of his father's deposition in June 1995. Far from being an anomaly in an Arab environment resistant to any change, the network and its distinctive 'politics of signification' is exemplary of the entirety of the regional political mutations originating from the first Gulf War.

These mutations demarcate the whole Arab satellite broadcasting landscape that instituted itself in the wake of the war as a new, deterritorialised political space where hegemonic and competitive strategies are played out. Financed and sheltered by the smallest Gulf monarchy, one of the richest and most free from interior social constraints (by reason of the weak demographics and the large foreign workforce), al-Jazeera has been utilised as a new tool of modern legitimisation at the hands of the new Qatari power elites. By creating a space for the public expression of a pan-Arabic demand for democracy and for the contest of imperial politics in the region, it fulfils a triple function. It is at the same time the best proof for the 'democratic' position of the al-Thanis, an efficient instrument with which to seek an autonomous regional role and an ingenious tool to cover up their strategic alliance with the US. This last function was certainly most obvious during the war on Iraq in 2003. In spite of the Qatari political and logistical involvement alongside the US in the conflict, al-Jazeera somehow managed to eclipse al-Siliyya, the command centre for the war on Iraq located only a few kilometres from the network's headquarters in Doha. By doing so, the Qatari regime paradoxically appeared more as promoter of an anti-US sentiment, in line with the Arab 'street', rather than supporting 'imperial goals'.

Pan-Arabism

During the decade of its existence, the journalistic narrative of al-Jazeera was articulated along two axes: democracy and pan-Arabism, the first contesting the politics of the Arab power elites and the second denouncing the politics of the US and Israel. These two repertoires, depending on variable combinations of the actuality and the conflicts in the region, were the distinctive feature of the network and allowed it to acquire legitimacy.

Yet, if this legitimacy seems indisputable with regard to the propensity of the network 'to opine',[28] notably in periods of crisis, it remains hard to measure in terms of links established with the public and the socialisation processes of which it is supposed to be the vector.

Nonetheless, it signifies the affirmation of a new player that resembles what Herbert Blumer calls 'a collective behaviour', animated by a strategy aimed at occupying a representational space claiming a new form of Arabity[29] and Islamity. This 'collective enterprise' is incarnated by a journalistic elite which crystallised as part of the network.[30] It is, in some ways, the media aspect of the phenomenon that Olivier Roy called the 'new pan-Arabism',[31] namely, a form of resurgence of the nationalist rhetoric leaning on simultaneously Islamic and democratic referents.

Breaking, since its genesis, with the Saudi model (based on the formula of a majority of Lebanese journalists and the bulk of the funding from Saudi),[32] al-Jazeera mobilised a pan-Arab personnel, addressing an Arabic public by constructing an actuality in Arabic which transcends borders and which investigates, above all, the 'seismic epicentres' of the regional tensions. The network thus provides an understanding of the conflicts that is simultaneously regional and endogenous. The Arabity of this elite manifests itself through the affirmation of 'an imaginary community, and imagined as much as intrinsically limited and sovereign',[33] contesting the US politics in the Arab world. In doing so, it strove to construct for its public a representation and a consciousness of itself, by making it its task to show its frustrations, its aspirations, a real or fantasised perception of exterior hostility, by revisiting its official historiography[34] and by drawing the contours of an otherness. Its pan-Arabism was equally expressed as a distancing asserted, vis-à-vis the western/exogenous journalistic narrative about the Arab world, and a determination to construct a global media pole catering to a diversified public (al-Jazeera International, al-Jazeera children's channel, al-Jazeera sports channel, al-Jazeera documentary channel, as well as a blog for the youth).

As to its Islamity, it rests on two pillars. The first is found in Youssef Al-Qaradawi. A well-known cleric of Egyptian origin, living in Qatar since the 1960s, president of the International Union of Muslim Ulema and displaying his distance from the Egyptian Muslim Brotherhood, he has appeared since the launch of al-Jazeera as the face of its religious credibility. Thereby benefiting from a weekly show (*Al-Shari'a wal-hayat*, the Sharia and everyday life), he presents an 'open Islam', opposed to both Salafi Wahabi (for example defending the right of women to vote) and traditionalist Islam (forbidding honour crimes). He shows himself as a *da'iya* intervening in worldly matters defending an Arab-Islamic identity, supporting the Palestinian struggle[35] and criticising the imperial politics of the US in the region. The second pillar is found in its treatment of political Islam. By providing Islamism with a visibility unequalled on US-financed networks, al-Jazeera inscribes itself as breaking with the dominant criminalisation of this phenomenon in the official media sphere. Its position was to account for this phenomenon, in its polysemy, by treating it like all other players, independent of its position in the institutional political landscape. In doing so, al-Jazeera did not hesitate to apprehend Islamism as a force of political change (the legislative elections in Palestine in 2006) or as national resistance (the role of Hizbullah in the 2006 war or the resistance in Iraq), nor as a belligerent party in an asymmetric war (al-Qa'ida during the war in Afghanistan).

Islamity is not the only cause of the renewal of this pan-Arabism. Al-Jazeera contributes, for its part, in three specific ways.

Above all, it is a pan-Arabism assisted by a technical and extraterritorialised mediation that is, as a result, at the heart 'of the tension between the tendencies towards individualisation and the process of socialisation'.[36] This new pan-Arabism, if it is part of the global phenomenon analysed by Olivier Roy, its social effects, its concrete articulation in the real world and the forms of identification that it evokes, remain difficult to apprehend. Thus, how to understand in the long run the effect of al-Jazeera on the political recomposition of the Islamist movements, of which most have gained – thanks to al-Jazeera – media exposure by getting around the censorship to which they are subject in their own countries? Or further, how to account for the consequences of the relentless examination of the Iraqi scene on the confessional tension in Iraq?[37]

Be that as it may, the distance established by technology, and accentuated by the presence of the network on the territory of a rich state not skimping on financing, affects this new pan-Arabism. The latter turns out to be sufficiently impermeable to social questions; rather, subscribing to the world's neoliberal approach. This can be seen in the coverage of the social movements that have developed in the Egyptian public sector since 2006, the representation of which showed that this type of conflict lies outside the network's categories of interpretation and its political culture.

Secondly, the position of the network is not the reflection of the ideology of the ruling elite. In spite of the ambiguous relationships that link al-Jazeera to the Qatari monarchy, conveyed by a certain indulgence, the network remains a non-state organ. Its components are more bonded by communal journalistic ethics (the opinion and its opposite), an avowed political identity and symbolic capital that is displayed (the prestige of the network). This leaves room for certain plasticity in the way events are narrated and a diversity of adopted formats of narration. Although intervening in a hierarchical framework ritualised and governed by bureaucratic logics of efficiency and management, and in spite of the diverse political constraints emanating from a number of power centres (the Qatari regime, the US and Arab regimes...) to which they are subjected, al-Jazeera journalists benefit from a real margin of autonomy, inherent in the symbolic weight of the station.

Nevertheless, this new pan-Arabism remains, in the end, a dependant of the Qatari regime; because ultimately it is its financial dependence that determines the contours of the network's horizon. The recent reconciliation between Qatar and Saudi Arabia, with the reestablishment of a Saudi diplomatic representation in Doha since 2002, motivated, it seems, by a common appreciation of the 'Iranian menace',[38] does not seem to have been without effect on al-Jazeera. For the first time in years, in January 2008, the network was authorised to cover the pilgrimage to Mecca; on the other hand, from then on it showed itself less impertinent in its coverage of the Saudi scene. Is this a circumstantial evolution? The answer will certainly not depend exclusively on the newsroom of the network, but principally on the evolution of the regional situation as perceived by the Qatari monarchy.

Is it therefore accurate to suggest that the network is a vector of pan-Arab renewal in the region as argued by Alterman, or of the version more contaminated by globalisation, the McArabism?[39] It is hardly possible to answer in a clear-cut manner. The conditions of the renewal of pan-Arabism are, in fact, indebted to the democratisation of the Arab world. The enchanted vision of the first studies, praising the network's role in this matter, is less and less present today in view of the durability of the authoritarian order in the Arab world. The immoderate recourse to the notion of the public sphere in the earlier analytic work[40] – seen as a major consequence of the arrival of al-Jazeera – seems to give way to investigation of its applicability and its ideologisation.[41] More and more researchers highlight, notably, the incapacity of one medium to cause, by itself, a democratisation of the political environment. Mohamed Zayani, for example, points out that 'Real change cannot be expected solely or mainly from the media sector. Democracy cannot emanate just from the media, the political systems and institutions themselves have to change, evolve, and adapt.'[42] Philip Seib states that in his opinion 'the information-influenced political climate is volatile as well as vibrant, making it hard to map the path toward constructive change.' Definitely, the ongoing evolution, the potency of which is hard to apprehend, does not seem unconnected to the extension of the realm of licit speech encouraged by the network. The same is the case for the consecration of political talk shows in the TV media or even the restructuring of the radio domain in the larger part of the Arab countries, which, in spite of its limits, signifies the end of the state monopoly over the radio. Evidently, our understanding of this evolution would profit from being refined by new anthropological research exploring the impact of technological and virtual mediation on real social relations.

Nevertheless, beyond even al-Jazeera, the boom of satellite news channels attests to the existence of a regional market for news that in turn bears witness to the existence of a cross-border Arab identity.[43] In this sense the different estimates and surveys confirm the primacy of the network as the first source of information, particularly in times of crisis, but this attests more to 'its capacity to define a horizon for debate and issues rather than Orwellian mind control'[44] – because the pan-Arabism of the network cannot act as a hypodermic needle on a public waiting for a call to mobilisation, even if certain public action seems to have been provoked by al-Jazeera.[45] The forms of social consumption and appropriation of this new pan-Arabism depend *de facto* on specifically when it is received. Certain recent events have shown that the latter is not always uniform. In this way, the Shi'i population in Iraq saw the fall of Saddam Hussein with a favourable eye. And more recently, during the shelling of the Palestinian camp Nahr al-Barid in northern Lebanon, the fate of the Palestin-

population attracted more attention from the al-Jazeera journalists than from the Lebanese public. Be that as it may, considering al-Jazeera a form of mediation of a new pan-Arabism allows – if not a grasp of its reception – an understanding of its specificity and complexity.

Notes

1. Anwar Sadat Chair for Peace and Development University of Maryland/Zogby International. 2006 Annual Arab Public Opinion. Survey. A six-country Study: Egypt, Jordan, Lebanon, Morocco, Saudi Arabia and UAE.
2. In the wake of an attack on a Jewish seminary in Jerusalem, the Israeli government decided to officially boycott the network for several days. This decision was justified by pointing to the tone of the coverage of the recent Israeli offensive in Gaza that was judged too 'militant'. *Le Monde*, 15 March, 2008.
3. Marc Lynch, 'Arab arguments: Talk Shows and The New Arab Public Sphere', pp. 101–118.
4. To use the term employed by Elias Sanbar. See Farouk Mardam-Bey and Elias Sanbar, *Etre arabe*, p. 246.
5. Interviews with management and journalists of the network conducted between 2003 and 2006 in Doha, Amman, Paris and Madrid.
6. Notably for its coverage of Operation Desert Fox under the Clinton administration.
7. Betting on the popularity of the network in order to offer it exclusive rights: visits to hospitals and the famous scenes of the Americans taken as prisoners of war.
8. The most representative episode was the complete re-broadcasting, without editing, of the images of the cadavers of soldiers and the interrogation of American prisoners of war aired by Iraqi TV. 23 March 2003.
9. Al-Jazeera sometimes fell for the official Iraqi propaganda. The bulletin aired on 24 March about the heroic exploits of an old Iraqi peasant who, supposedly, had single-handedly brought down an Apache combat helicopter with his old rifle is worth remembering.
10. Philippe Droz-Vincent, *Vertiges de la puissance. Le 'moment américain' au Moyen-Orient.*
11. Stuart Hall, *Identités et culture. Politiques des Cultural studies.*
12. In his 2004 State of the Union address, the head of the White House did not hesitate to denounce its responsibility in the 'hateful propaganda' of which the sole superpower would become the target.

13. Olfa Lamloum, *Al-Jazira miroir rebelle et ambigu du monde arabe*, pp. 127–130.
14. Michael Massing, 'Now, tell us'.
15. In May 2003, Mohamed Jassem al-Ali is relieved from his post following American accusations of collaboration with the intelligence services of S. Hussein.
16. Looking closer, in January 2008 for example, the programme treating Iraq (al-mashhad al-'iraqi) dedicated its four episodes to political actors supported by the US.
17. Olivier Hahn, 'Cultures of TV News journalism and Prospects for a transcultural public Sphere', pp. 13–27.
18. Term used by President Bush during his State of the Union address in 2004.
19. Naomi Sakr, *Arab Media and Political Renewal. Community, Legitimacy and public life*.
20. Al-Quds Al-Arabi reports that the meeting of the Arab ministers of information estimated the number of satellite channels at about 500. Al-Quds Al-Arabi 14 February 2008.
21. To adopt the expression of the Syrian intellectual Sadik Jalel Al-Azm, *Siyyâssât Kartir wa munadhirî al-hiqba al-sa'udiyya*.
22. Fawaz Traboulsi, *Dhofar, shahada min zaman al-thawra*.
23. See Ghassan Salame, *Al-Siyyâssa al-kharijiyya al-sa'udiyya munthu 'âm 1945. Dirâsa fî al-'alâqât al-dawliyya*.
24. A number of startups of satellite stations are launched by members of the Saudi royal family: Middle East Broadcasting Corporation (MBC) in 1991, Arab Radio and Television (ART) and Orbit Communication in 1994.
25. Bruce Riedel and Bilal Y. Saab, 'Al-Qa'ida 's Third Front: Saudi Arabia', pp. 33–46.
26. Evidently without questioning the strategic alliance sealed in 1945 between ibn Saoud and Roosevelt, based on the exchange of 'security against oil'.
27. Marc Valeri, 'Réveil laborieux pour l'Etat-Qabous. Identité nationale et légitimité politique dans l'Oman d'aujourd'hui', pp. 37–57
28. To take up the proposition of Erik Neveu, 'Les sciences sociales face à l'espace public, les sciences sociales dans l'espace public', p. 42.
29. In the sense developed by Maxime Rodinson.
30. Here a fact that cannot go by without remembering the famous remark of Acton: 'exile is the breeding-ground of nationalism.'
31. Olivier Roy, 'L'islamisme, nouveau panarabisme'.
32. At the origins of the foundation of the network is a group of fifteen journalists formerly with the BBC Arabic service.
33. To take up the definition of Benedict Anderson, *L'imaginaire national. Réflexions sur l'origine et l'essor du nationalisme*.

34. The programme Shâhid 'ala al-'asr (witness of the century) hosted by Ahmad Mansour has been stopped for some time and is an example of this.
35. Including the Hamas leadership.
36. Pierre Chambat, 'Espace public, espace privé: le rôle de la médiation technique', p. 69.
37. The network has regularly been accused for its pro-Sunni position in Iraq, notably by Shia ulemas in Saudi Arabia. The last remarkable incident goes back to May 2007. Ahmed Mansour, journalist and presenter of the programme 'without frontiers' was accused of denigrating the Shia marja'iyya of Najaf (Sistani). For the text of the Shia ulemas see < http://www.saudinfocus.com/ar/forum/archive/index.php?t-34143.html>.
38. Robert F. Worth, 'Al-Jazeera No Longer Nips at Saoudis', *New York Times*, 4 January, 2008
39. See Khalil Rinawi, *Instant Nationalism. Mc Arabism, Al-Jazeera and Transnational Media in the Arab World*, p. XIII. He defines McArabism as 'a situation in which citizens throughout the Arab World receive identical nationalist pan-Arab content via transnational media, just as one can get the same big Mac at any McDonald's outlet.'
40. Eicklman, Dale F. and Jon W. Anderson, *New Media in the Muslim World: the Emerging Public Sphere*.
41. Franck Mermier, 'Médias et espace public pan-arabe: de quoi parle-t-on?'.
42. Mohamed Zayani, 'introduction. Al-Jazeera and the Vicissitudes the New Arab Mediascape', p. 35.
43. J. B. Alterman, 'Counting nodes and counting noses: understanding new media in the Middle East'.
44. Erik Neveu, *Sociologie du journalisme*, p. 84.
45. Notably the demonstration in solidarity with the Palestinians of Jenin that took place in certain Arab countries.

Bibliography

Al-Azm, Sadik, *Siyyâssât Carter wa munadhirî al-hiqba al-sa'udiyya*, Beyrouth: Dâr al-tali'a, 1977.

Alterman, J. B., Counting nodes and counting noses: understanding new media in the Middle East, *The Middle East Journal*, summer, 2000, vol.54, n° 3.

Anderson, Benedict, *L'imaginaire national. Réflexions sur l'origine et l'essor du nationalisme*, Paris: La Découverte, 1996.

Annual Arab Public Opinion. Survey. A Six-Country Study: Egypt, Jordan, Lebanon, Morocco, Saudi Arabia and UAE, Anwar Sadat Chair for Peace and Development University of Maryland/Zogby International. 2006.

Blumer, Herbert, 'Collective Behavior', in A.M. Lee, ed., in *New outline of the principles of sociology*, New York: Barnes and Noble, 1946, pp. 166–222.
Chambat, Pierre, 'Espace public, espace privé : le rôle de la médiation technique', in Pailliart, Isabelle, ed., *l'espace public et l'emprise de la communication*, Grenoble: Ellug, 1995, pp. 65–98.
Droz-Vincent, Philippe, *Vertige de la puissance. Le moment américain au Moyen-Orient*, Paris: La Découverte, 2007.
Eickelman, Dale F. and Jon W. Anderson, *New Media in the Muslim World : The Emerging Public Sphere*, Bloomington: Indiana University Press, 1999.
Hahn, Olivier, 'Cultures of TV News Journalism and Prospects for a Transcultural Public Sphere', in Sakr, Naomi, ed., *Arab Media and Political Renewal. Community, Legitimacy and public life*, London: I. B. Tauris, 2007, pp. 13–27.
Hall, Stuart, *Identités et cultures. Politiques des cultural studies*, Paris: Editions Amsterdam, 2007.
Lamloum, Olfa, *Al-Jazira miroir rebelle et ambigu du monde arabe*, Paris: La Découverte, 2004.
Lynch, Marc, *Voices of the New Arab Public. Iraq, al-Jazeera, and Middle East Politics Today*, New York: Columbia University Press, 2006.
Maigret, Eric and Eric Macé, *Penser les médiacultures. Nouvelles pratiques et nouvelles approches de la représentation du monde*, Ina: Armand Colin, 2005.
Mardam-Bey, Farouk and Elias Sanbar, *Etre arabe*, Paris: Actes Sud/Sindbad, 2005.
Massing, Michael, 'Now, tell us', *New York Review of Books*, September 2004.
Mermier, Franck, 'Médias et espace public pan-arabe: de quoi parle-t-on?', à paraître 2008.
Najm Jarrah, 'The rise and decline of London as a pan-Arab media Hub', *Arab Media and Society*, January 2008 <http://www.arabmediasociety.com/?article=571>
Neveu, Erik, *Sociologie du journalisme*, Paris: La Découverte, 2004.
Neveu, Erik, 'Les sciences sociales face à l'espace public, les sciences sociales dans l'espace public', in Pailliart, Isabelle, ed., *l'espace public et l'emprise de la communication*, Grenoble: Ellug, 1995, pp. 37–64.
Riedel, Bruce and Bilal Y. Saab, 'Al-Qa'ida 's third Front: Saudi Arabia', *The Washington Quarterly*, spring 2008, pp. 33–46.
Rinnawi, Khalil, *Instant Nationalism. Mc Arabism, Al-Jazeera and Transnational Media in the Arab World*, University Press of America, 2006.
Rodinson, Maxime, *Les Arabes*. Paris: Puf, 1979.
Roy, Olivier, L'islamisme, nouveau panarabisme, 2006. <http///www.ceri-sciences-po.org>
Sakr, Naomi, 'Channels of Interaction. The role of Gulf-Owned Media firms in Globlisation', in Paul Dresh and James Piscatori, eds, *Monarchies and Nations.*

Globalisation and Identity in the Arab States of the Gulf, London: I. B. Tauris, 2005, pp. 34–51.

Sakr, Naomi, ed., *Arab Media and Political Renewal. Community, Legitimacy and public life*, London: I. B. Tauris, 2007.

Salame, Ghassan, *Al-Siyyâssa al-kharijiyya al-sa'udiyya munthu 'âm 1945. Dirâsa fî al-'alâqât al-dawliyya*, Bayrout:Ma'had al-inm'a al-'arabî, 1980.

Seib, Philip, *New Media and the New Middle East*, New York: Palgrave Macmillan, 2007.

Traboulsi, Fawaz, *Dhofar, shahada min zaman al-thawra* (Dhofar, a testimony from Revolution Days), Beirut: Riad al-rayyes Books, 2004.

Valeri, Marc, 'Réveil laborieux pour l'Etat-Qabous. Identité nationale et légitimité politique dans l'Oman d'aujourd'hui', in Maghreb-Machrek, n°177, automne 2003, Paris, pp. 37–57.

Zayani, Mohamed, 'Introduction. Al-Jazeera and the Vicissitudes of the New Arab Mediascape' in Mohamed Zayani, ed., *The Al-Jazeera Phenomenon: Critical Perspectives on New Arab Media*, Boulder, Co: Paradigm, 2005.

LAWRENCE PINTAK

Journalist as Change Agent –
Government repression, corporate feudalism and the evolving mission of Arab journalism

Change. It was the one constant of Arab journalism in the first decade of the twenty-first century. Change in the structures of the region's media. Change in the relationship between journalists and governments. Change in the very role played by Arab journalists as they struggled to redefine themselves and their mission.

Saudi editor Othman al-Sini put his finger on the essential question being asked in many newsrooms: 'I wonder if media should be change-makers or reporters of change?'[1]

For the greybeards of Arab journalism, old enough to have covered the Arab-Israeli wars or Sadat's visit to Jerusalem, the mediascape in which they now operated was almost unrecognisable compared with the one in which they learned their craft. No longer did governments have a monopoly on truth; no longer could officials be sure their dirt would be hidden; no longer could elaborate propaganda fantasies masquerade as fact.

It was unthinkable that a spectacular battlefield defeat could be sold to the public as a masterly victory without the official story instantly being undermined, as Egypt managed during the war in the Sinai. It was unthinkable that a regime could crush an internal uprising without images of the slaughter slipping out, as Syria managed in the city of Hama during the 1980s. It was unthinkable that an Arab country could be invaded and citizens of its neighbor remain oblivious to that fact because their media remained silent about the news, as was the case with the Saudis for the first forty-eight hours after Iraq's invasion of Kuwait.

If any one factor could be given credit for these changes, it was the Arab television revolution. By 2008, there were more than 300 free-to-air Arab satellite channels, on a spectrum once entirely controlled by governments. A cacophony of opinions now replaced the godlike pronouncements of the state-run channels. At the same time, the so-called 'al-Jazeera Effect' had changed the way journalists saw themselves.[2]

This writer is a journalistic child of the Watergate generation, part of the wave of US students entering journalism schools in the early 1970s, inspired by the crusading journalism of Bob Woodward and Carl Bernstein of the *Washington Post*, who helped bring down Richard Nixon. We all wanted to change the world. Thirty years later, young Arab journalists could be heard saying much the same.

'I can't criticise from *within* my country,' wrote one of my students at The American University in Cairo, explaining why she wanted to report for the Arab satellite channels, 'but journalism allows me to criticise from *outside* and begin to make things different.' That sense of the possible could also be found among some of those greybeards mentioned above; men and women who, as young reporters, were embarrassed by the mouthpiece nature of their profession but, inspired by the new style of journalism pioneered by al-Jazeera, were themselves pushing the envelope.

But the regimes at the receiving end of this form of critical journalism have not been oblivious to the sea change in the media-government relationship. Al-Jazeera itself was the first target of their wrath. At one time or another, its reporters have been thrown out of virtually every Arab country. Journalists in every medium in every Arab country have also felt the sting. From Morocco to Yemen, Syria to Sudan, journalists have been threatened, harassed, jailed and killed. Iraq has become the graveyard of journalists, with more media workers losing their lives than in any conflict since World War 2. They have fallen at the hands of US forces, the Iraqi military and militants of every religious and ideological stripe. Arab journalists have paid the greatest price, both those working for Arab news organisations and those employed by Western news organisations, going where American and European journalists fear to tread.

But the deaths of journalists are not confined to Iraq. The brutal assassinations of publisher Gibran Tueni and columnist Samir Kassir of Lebanon's leading newspaper, *an-Nahar*, were the most high-profile evidence of that. Both were outspoken critics of the Syrian military presence in Lebanon. And May Chidiac, a programme host on the Lebanese Broadcasting Corporation, who lost an arm and a leg when a bomb wired to her ignition blew up, was a living reminder of the dangers of criticising Arab regimes. The graves of journalists who challenged the system could also be found in many other Arab countries, while others rotted in jail cells for trying to practise their craft. The sea of black on the press freedom maps published by Freedom House told the story. Not a single Arab country garnered a 'free' designation and just two – Kuwait and Lebanon – were categorised as 'partly free'.[3] Egypt, named by Reporters sans Frontiers as one of its 'predators of press freedom',[4] was but one example. In just three months in the fall of 2007, seven reporters and editors were sentenced to forced labour for 'harming the reputation of the state' or its legal system – code for criticising the president and his cronies.

The grim irony was that journalists were figuratively and literally under the gun precisely because things were changing. Many were inspired to take risks they would have never contemplated a decade before. Indeed, it was not until the very end of the twentieth century that the very media through which much of this new, aggressive form of reporting – satellite TV, blogs and new semi-independent newspapers – began to appear.

'Without question, literate Arab citizens today are more connected to global media flows and have better access to information about their own countries and the world than any previous generation,' according to a Brookings Institution study, while '[a]cross the region, a new generation of presidents and kings stress their openness to new media and their embrace of the Internet.'[5]

Yet that embrace was often accompanied by a knife in the back of media reform. Governments went through the motions of passing new laws and regulations that were trumpeted as guarantees of freedom of expression, yet such public acts were often part of an elaborate smoke-and-mirrors game that veiled the reality of continued draconian regulation. As the Committee to Protect Journalists (CPJ) reported, 'Image conscious governments have also become masters of spin, championing cosmetic media reforms designed mainly for public consumption.'⁶ For example, on his visits to the West, Jordan's King Abdullah II was a voluble proponent of press freedom; back at home, the passage of new press regulations that dropped criminal penalties for reporters masked the fact that it was for alleged violations of a host of other laws that journalists were being sent to jail. 'As long as you don't write about the king, the military, religion or sex you can cover anything you want,' was how Jordanian journalist Sameh al-Mahariq summed up the state of affairs.⁷

Truth was no defence; no charge too absurd. In the UAE, a prison sentence against a website editor remained in force even though the libel suit that prompted the conviction was withdrawn. An Egyptian editor faced years of forced labour because an Egyptian government loyalist said he had suffered 'emotional damage' when he read a story that President Mubarak might be ill or dead.

The challenge of objectivity

Poor training, low pay and a preponderance of opinion are the legacies of the Arab media's history as a propaganda tool. With all broadcast outlets in the hands of governments, the term 'television journalism' was an oxymoron. Meanwhile, it said much about the nature of professionalism in the industry that Muhammad Hassanein Heikel, confidant of, and mouthpiece for, Egyptian dictator Gamal Abdul Nasser, was widely admired as the embodiment of Arab journalism.

As they seek to shake off that legacy, Arab journalists are faced with a host of questions about their role and their relationship to the state, to the forces of domestic opposition shaking the region and to the presence of Western armies in the Arab heartland. At the heart of this set of questions lies the elusive concept of journalistic objectivity. It is the Holy Grail of many Western journalists, but a prize rarely, if ever, captured. Media historian Michael Schudson defines objectivity as 'the view that one can and should separate facts from values.'⁸ It is the polar opposite of the Islamic view that 'information and knowledge are not value-free, but have normative, ethical and moral imperatives.'⁹

Religion aside, objectivity is a normative ideal that often dissolves in the face of the grim realities journalists face in their professional lives. The degree to which US journalists jettisoned any pretence of objectivity in the wake of 9/11 was epitomised by the comments of Dan Rather, then anchor of *The CBS Evening News*, a few days after the attacks. With tears in his eyes, he told a television interviewer, 'George Bush is the president. He makes the decisions and ... wherever he wants me to line up, just tell me where.'[10] For Arab journalists, faced with the daily spectre of fellow Arabs killed as a result of the US invasion of Iraq or Israeli military actions in the Occupied Territories, there arose the question of whether discussion of some elusive ideal of objectivity was even appropriate. They were Arabs, and Arabs were dying. What was there to be objective about? For those Arab journalists on the frontlines, like Ahmed Mansour, al-Jazeera's correspondent in the Iraqi city of Fallujah during the siege, theory seemed a far cry from reality:

> A lot of times we can [make] a gap between being a human being and being a reporter, but sometimes you cannot. And when I was in Fallujah, every girl I saw, I remembered my daughter. When I try to [separate myself from what was happening], sometimes I cannot. I saw a child injured or die, I remember my son. They are Arab like you, Muslim like you. They are human people like you. Sometimes my colleagues [at] al-Jazeera tell me, 'Ahmed, you can talk more quiet and more slowly.' How can I when this plane over my head destroys everything? Sometimes when they contacted me, I was in the middle of a massacre. How can I talk quietly? Sometimes I was on the air from in the hospital and between bodies and people injured. You are a man in the end. You are not an angel or from another planet. You are from this area. So you cannot do this gap between you as journalist and you as human.[11]

Bush administration criticism of alleged Arab media bias ignored a fundamental fact: Arab journalists were reporting from an *Arab* perspective. Arab reporters were inside Fallujah, while US journalists were outside, embedded with the Marines. Arab journalists were inside Ramallah and Jenin during the Israeli sieges of those West Bank cities, Western journalists were outside with the Israelis. The point of view could not have been more different. Western journalists, to a large degree, witnessed the conflicts from the perspective of the besieging troops; Arab journalists saw the bloody consequences on the ground. No matter how Arab – or

Western – journalists saw their role, culture and place affected how they saw the stories they covered. They framed their stories – in terms of both narrative and physical cropping of pictures – through a different lens. If, as some scholars argue, journalism is 'a set of cultural practices'[12] and journalism itself is an 'ideology,'[13] those practices, and that ideology, are themselves shaped by the environment in which they are practised, just as Communism differed in the Soviet Union and China and the Judeo-Christian or Islamic cultures find many forms around the world. 'The way to select news and the way to select [quotes], the way to select the point of view, this still belongs to our cultural environment and background,' argued al-Jazeera anchor M'hamed Krichen. 'We cannot separate ourselves from this background, like the American or French or British journalists cannot do the same.'[14]

One man's terrorist was, of course, another man's freedom fighter. And one nation's campaign to 'free' the Iraqi people was another nation's invasion. The pictures told the story; America television viewers saw the missiles being launched, Arab viewers saw the remains of the people upon whom they fell. To Americans, the iconic image of Saddam's statue being toppled spoke of the fall of a dictator. For Arabs, the image was figuratively and literally framed differently; the photo was cropped to show the US troops who pulled it down and the narrative spoke of colonial occupation.

'The Arab journalist, in general, has the same problem as the average Arab,' explained Mansour's al-Jazeera colleague, Samir Khader, 'this feeling of being under siege. And they have a mission to defend the nation, to defend the religion.'[15] Khader's view reflects the broad sentiment among Arab journalists as evident in a survey of 601 Arab reporters and editors a colleague and I carried out in fourteen Arab countries.[16] The journalists gave high priority to a group of journalistic roles that we labelled the Guardian typology. It included such functions as supporting the Palestinian cause, fostering Arab culture, encouraging spiritual values, defending Arab interest, enhancing pan-Arab unity and protecting Islamic tradition. Such a worldview should be of no surprise. Journalists cannot be separated from the environment and culture in which they operate, and the 'rally-round-the-flag' effect[17] is a common function of media around the world in times of crisis. A state of siege existed in the Arab region and broader Muslim world post-9/11, a place where inspiring the masses in defence of the homeland against real or imagined threats was a traditional role of government media.

But among Arab journalists, defence of the homeland still falls second to reform of the homeland. While concerned about the external threat, Arab journalists give top priority to political and social change within Arab society. Fully 75 per cent of respondents to our survey said 'political reform' was the most important job of a journalist, while large percentages supported a set of roles, which include using news for the social good, encouraging civil engagement and serving as a voice for the poor and transforming society, that together embody the functions of a Change Agent.

'The role of the media in shaping public debate and covering politics is one of the most important concerns facing journalists today,' according to Saudi journalist Samar Fatany. '[It is] [o]ur mission ... to mobilise and move the public debate toward positive attitudes and global thinking – dire needs for the progress and development of our country.'[18]

Corporate feudalism

In countries emerging from a system of centralised control, a pattern of media evolution is often witnessed. When the restrictions on media ownership are lifted, there is often an explosion of new media outlets resulting in a free-for-all in the marketplace. In Indonesia, for example, the number of media outlets and the number of journalists quadrupled in the five years after the overthrow of President Suharto, who had ruled with an iron grip for three decades.[19] That explosion is soon followed by a reality check, as the battle for a limited pool of advertising revenues results in one of several outcomes: Many of the upstarts go bankrupt, others sell their souls, accepting under-the-table funding from political or business groups in return for positive coverage, and a third group is bought up by media conglomerates that often emerge. In Russia, media eventually went full circle, transitioning from Soviet state control to independence, after which they were bought up by the emerging Russian oligarchs, who were themselves then seen as a threat by the Putin regime, which crushed the oligarchs and brought media back under the government's thumb.

The Arab world is in danger of skipping those intermediate steps, transitioning directly from state control to a system of corporate feudalism, in which media properties are owned by members of ruling families, individuals close to the centers of power, or would-be politicos of various ideological and religious stripes. The largest media conglomerate in the Arab world is controlled by the brother-in-law of King Abdallah of Saudi Arabia; Qatar's royal family controls al-Jazeera; the late King Faisal of Saudi Arabia owned 39 per cent of the shares in *al-Sharq al-Awsat*; Libya's media properties are all in the hands of Muammar Qadaffi's son, Seif; the son-in-law of Tunisian President Zine al-Abdine Ben Ali won the licence for a new radio station sought by numerous business groups and the country's only private TV station is owned by a pro-regime businessman; Lebanon's Future TV is owned by – and serves as a cheerleader for – the son of the slain former prime minister, Rafiq Hariri; Egypt's newest family of channels is being built by a powerful tycoon who controls a construction and telecommunications empire closely linked to the Mubarak regime; and on it goes.

'Arab rulers, regardless of their differences, agree on one thing – all of them consider the Arab press to be their sworn enemy,' according to Jamal Amer, editor-in-chief of *al-Wasat* in Yemen.[20] No one was free; everyone operated within 'red lines' that defined areas too hot to touch. For all-news al-Arabiya, and other members of the Saudi-owned MBC Group, that included internal Saudi politics, terrorism and even religion in the nation that housed the city to which every Muslim in the world prays five times a day. For al-Jazeera, red lines involved issues that affected Qatari foreign relations, evident in the suddenly softening of its coverage of Saudi Arabia in late 2007, which sources in the newsroom confirmed to this writer was the result of a Qatari-Saudi rapprochement in the face of the perceived threat from Iran. For many journalists writing for domestic papers in Arab countries, those red lines were firmly etched around the palaces of the ruling elite.

An example of the cosy relationship between Arab media moguls and the rulers themselves came from the violence-plagued Sudanese province of Darfur, where a young director for the MBC Group produced a documentary called *Jihad on Horseback,* which documented the Sudanese government's complicity in the slaughter. A phone call from Sudanese President Omar al-Bashir to the then-crown prince of Saudi Arabia, Prince Abdallah, was all it took to kill the programme.[21]

Islam, nationalism and the issue of identity

The relationship between Islamism, Arab nationalism and political liberalism is complex. At times they have been warring ideologies; at other points in history, uneasy allies. For Islamists, the long-term aim is the creation of an Islamic society that stretches from Europe to Southeast Asia. The traditional goal of Arab nationalists was the unification of the Arab homeland. Political liberals seek some form of participatory democracy. In the post-war period, Islamism and Arab nationalism shared the short-term goal of ridding the region of its colonial masters. In the 1950s, the Islamists of the Muslim Brotherhood joined with Gamal Abdul Nasser and his Free Officers to overthrow the monarchy. But in each case, the alliance of convenience soon ended.

There is today a similar confluence of interests as Islamists, Arab nationalists and political liberals seek political and social change in the region. The result is a new Arabism, borne of an 'emotional unity' or *wihdathal* (unity of situation) in the face of internal and external threats[22] and fuelled by the rise of satellite television, which has created an experience shared by all Arabs.

One vivid manifestation of that could be seen in the demonstrations in Cairo at the end of the 2006 Lebanon war, where the crowds carried posters of the leader of the Shi'i Hizbullah movement, Hassan Nasrallah, alongside those of the godfather of Arab nationalism, Gamal Abdul Nasser – all carried live on satellite TV.

Make no mistake, Islamism, Arabism and those who support democratic change frequently came to blows, figuratively and literally. That could be seen on the confrontational talk shows on satellite television, in the newspapers, and in violence from the slums of Algiers and the hills of Yemen. It could also be felt in many of the newsrooms of the Arab world, as reporters and editors jockeyed over what stories to cover and how they should be framed. Nevertheless, journalists were at the forefront of the rise of this new Arabism. In our survey, we found that fully one third of Arab journalists identified first with the Arab world, a quarter with the Muslim world and just 15 per cent with the nation in which they were born. Politically, half identified themselves as 'democrat,' and, in a statement that said much about the evolution of the profession, when given a list that included

'Arab,' 'Muslim,' their nationality, and 'journalist,' and asked their primary identity, fully 50 per cent chose 'journalist.'[23]

M'Hamed Krichen of al-Jazeera summed up this new worldview. Asked to identify himself, he replied, 'I am an Arab, Muslim journalist.' In the newsrooms of the Arab world, at least, the old fault lines were beginning to blur.

The question of loyalty

As the sands shift under the Arab media, journalists are asking many questions of themselves. Some of those questions came in the least likely of places. 'To whom should loyalty be?' Abd el-Latif el-Menawy, head of news at state-run Egypt television, wondered aloud. 'Should the loyalty be to the receiver of information, whether it is a reader or a viewer? Or should the loyalty be to the society? Or should the loyalty be to the one who is paying the salary at the end of the month?'[24] A few years before, for the government television channel in the largest Arab country, the answer would have been self-evident.

Inseparable from that question of loyalty was the whole issue of mission and the seeming disconnect between the erstwhile norms of Western journalism and the realities of the Middle East in the early twenty-first century, something not likely to be resolved any time soon.

As for Saudi editor al-Sini's question about whether journalists should be change-makers or reporters of change, in many ways, Arab journalists found a way to combine both. As noted above, many Arab journalists saw it as their overt duty to drive political and social change. But whether or not all Arab journalists shared that view, the reality is that Arab media was *driving* change through the very fact that it was *reporting* change. Live, 24/7 coverage of anti-Syrian demonstrations in Beirut both helped protect the crowds from the kind of military backlash common in the years before satellite television and inspired movements for change in other parts of the Arab world. Police torture videos posted on YouTube by Egyptian bloggers and picked up by the mainstream media ultimately led to the unprecedented conviction of the policemen involved. News reports about a Saudi woman who was gang raped and then sentenced to prison and flogging for having illegal sex eventually prompted the government to overturn the sentence. These are but a few examples of the effect the new Arab journalism is having across the region.

As I have written elsewhere, television alone cannot create democracy, but it does act as a powerful tool for those who seek change.[25] The best reporting shines a light on injustice and provides a balanced view of events; two characteristics long absent from the Arab media. Therefore, however they see themselves – change-makers or reporters of change – Arab journalists who had shed the old ways were, by definition, helping to create change, in their industry and in their world.

Notes

1. Othman Mahmoud al-Sini, personal interview, Dubai: 2006.
2. Pintak, *Reflections in a Bloodshot Lens: America, Islam and the War of Ideas.*
3. 'Press Freedom Findings in Middle East and North Africa Disappointing after Earlier Hopeful Years', New York: Freedom House, 2007.
4. 'Predators of Press Freedom', Paris: Reporters Sans Frontières, 2007.
5. Heydemann, 'Upgrading Authoritarianism in the Arab World', pp. 19–20.
6. Joel Campagna, *Attacks on the Press 2007: Middle East and North Africa*, Committee to Protect Journalists, 2007, available from [http://www.cpj.org/attacks07/mideast07/mideast_analysis_07.html], (cited 3 February 2008).
7. Quoted in John Smock, 'Final Report, Knight International Press Fellow', Washington, DC: International Center for Journalists, 2006.
8. Schudson, *Origins of the Ideal of Objectivity in the Professions: Studies in the History of American Journalism and American Law, 1830–1940*, p. 3.
9. Mowlana, 'The New Global Order and Cultural Ecology', p. 13.
10. *The Late Show with David Letterman*, [television] (CBS, 17 September 2001 [cited 3 November 2003]); available from [http://www.mediaresearch.org/cyberalerts/2001/cyb20010918.asp].
11. Ahmed Mansour, personal interview, Doha: 2006.
12. Breen, *Journalism: Theory and Practice.*
13. Deuze, 'What Is Journalism? Professional Identity and Ideology of Journalists Reconsidered'.
14. M'Hamed Krichen, personal interview, Doha: 2005.
15. Samir Khader, personal interview, Doha: 2005.
16. Pintak and Ginges, 'The Mission of Arab Journalism: Creating Change in a Time of Turmoil'.

17. Mueller, 'Presidential Popularity from Truman to Johnson'.
18. Samar Fatany, 'Journalists Face Challenges Preserving Watchdog Role', *Arab News*, 12 February 2007, [cited Mar 5 2007]; available from http://www.arabnews.com/?page=7&section=0&article=92074&d=12&m=2&y=2007].
19. Harsono, 'Journalists Confront New Pressures in Indonesia'.
20. *Media Forum Examines Widespread Press Freedom Abuses*, World Association of Newspapers, 11 December 2006 [11 December 2006]; available from [http://www.wan-press.org/article12549.html].
21. Pintak, 'Nabil Kassem on Arab Media Darfur Silence' (Audio Interview), *Arab Media & Society*, 25 April 2007 [cited 2007]; available from [http://arabmediasociety.sqgd.co.uk/audio/index.php?item=17].
22. Abu Khalil, 'A New Arab Ideology? The Rejuvenation of Arab Nationalism', p. 46.
23. Pintak and Ginges, 'Mission of Arab Journalism'.
24. el-Menawy, 'Comments at Arab Broadcast Forum'.
25. Pintak, 'Satellite TV and Arab Democracy'.

Bibliography

Abu Khalil, As'ad, 'A New Arab Ideology? The Rejuvenation of Arab Nationalism', *The Middle East Journal*, no. 46 (1), 1992.

Breen, Myles, *Journalism: Theory and Practice*, Paddington, NSW, Australia: Macleay Press, 1998.

Deuze, Mark, 'What Is Journalism? Professional Identity and Ideology of Journalists Reconsidered', *Journalism*, no. 6 (4), 2005.

El-Menawy, Abd el-Latif, 'Comments at Arab Broadcast Forum', Abu Dhabi: 2006.

Harsono, Andreas, 'Journalists Confront New Pressures in Indonesia', *Neiman Reports*, 2002.

Heydemann, Steven, *'Upgrading Authoritarianism in the Arab World'*, Washington, DC: The Brookings Institution, 2007.

Mowlana, Hamid, 'The New Global Order and Cultural Ecology', *Media, Culture and Society*, no. 15 (1), 1993.

Mueller, John E., 'Presidential Popularity from Truman to Johnson', *The American Political Science Review*, no. 64 (1), 1970.

Pintak, Lawrence, *Reflections in a Bloodshot Lens: America, Islam and the War of Ideas*, London: Ann Arbor, MI: Pluto, 2006.

Pintak, Lawrence, 'Satellite TV and Arab Democracy', *Journalism Practice 2*, no. 1, 2008.

Pintak, Lawrence and Ginges, Jeremy, 'The Mission of Arab Journalism: Creating

Change in a Time of Turmoil', *The Harvard International Journal of Press/Politics*, no. 13 (3), 2008.

Schudson, Michael, *Origins of the Ideal of Objectivity in the Professions: Studies in the History of American Journalism and American Law, 1830–1940*, New York: Garland Pub, 1990.

Smock, John, *'Final Report, Knight International Press Fellow'*, Washington, DC: International Center for Journalists, 2006.

PART TWO

Market and Censorship

NAOMI SAKR

Is the Arab TV Viewer a King or a Pawn? How Arab Broadcasters Deal with Schedules and Audience Data[1]

The mention of audiences was for a long time conspicuous by its absence from face-to-face discussions about changes in Arab television. While one media forum after another – in Doha, Dubai or Beirut[2] – tackled issues around the suitability of television content and the ethics of television professionals, the word 'audience' rarely crept into formal presentations. When it did, it was usually embedded in a speaker's assumption that he or she, in their own capacity as a television viewer, had intimate knowledge of what everyone else on the receiving end of decisions about programme production and distribution would say about television programming if given the chance. Today, however, after nearly two decades in which Arab satellite channels have mushroomed with increasing velocity, there seems to be growing evidence that television audiences may, at long last, be getting at least some of the attention they deserve. Surveys of media use among Arab populations are reported with greater frequency and based on different samples from before. A myriad of small, specialised channels have emerged to tap into viewers' niche interests in anything from old movies to cooking to horse racing. The big networks have acknowledged that viewers appreciate innovation and creativity, while some media analysts have even gone so far as to report a shift in the balance of power from channel owners to viewers. 'The viewer has become king' is the way one explained it in early 2008.³

For the viewer to 'become king' would represent a rise in status that is remarkable enough to merit closer scrutiny, in order to gauge the accuracy of the analogy in the Arab context, based on a brief exploration of some recent developments in audience measurement, opinion polling and programming choices.⁴ It begins by considering where television viewers generally stand, whether they are inside or outside the Arab world, in a television industry pecking order that also includes owners, managers and advertisers. By setting the scene in this way it aims to provides pointers as to the kinds of development in ratings research and scheduling that could reliably indicate how much power audiences in Arab countries really have. Do audiences determine what television programmes are made available for them to watch or do they have to content themselves with whatever broadcasters decide they will get?

Citizen or consumer? Public or market?

Different media producers perceive their audiences in different ways. A broadcaster who relies on advertising revenue to pay for the making of programmes is usually obliged to think of viewers as potential consumers of items that are advertised in programme breaks. Audiences in this context come to be seen in terms of demographics and as a commodity in their own right, in the sense that advertisers are ready to buy access to audiences from broadcasters. Primary features determining their attractiveness as a market for advertisements, and thus for programmes, include incomes and discretionary control over personal and household spending. These tend to be linked to age, gender, education and lifestyle. Where awareness of such characteristics influences the output of media texts, as in the case of consumer-oriented programmes about fashion, cooking, travel, real estate and so on, it can be argued that those texts in turn construct the audience in a way that reinforces some of their features but ignores others. Segmenting audiences in this way is good for business, because it helps producers to identify and target those groups that will either maximise current revenue or bring the greatest future revenue growth.⁵ For the audience, in contrast, the implications may be less beneficial. Segmentation means that, out of the vast potential range of attributes and identities around which audiences might coalesce if given the opportunity, programmes actually on offer group them on the basis of only a limited range of attributes and identities, while at the same time encouraging audiences to perceive themselves in those same limited terms.[6]

It is possible, on the other hand, to perceive the television audience as a public rather than a market. The distinction here is between consumption and citizenship. Whereas advertising-funded commercial broadcasting constructs citizens primarily as consumers, there is an alternative model, which brings the viewers' role as citizens to the fore.[7] The latter is usually, but not exclusively, characteristic of public service television. In the world of Arab satellite channels, broadcasters occasionally reveal whether they see their channels as serving a market or a public. For some the stated difference is in their particular target market. For others the difference is between market and public. Orbit, a Saudi-owned pay-TV network launched in the mid-1990s, declared in its promotional literature that it would aim its programmes at 'well-travelled and affluent professionals'. Orbit's first chief executive, Alexander Zilo, said the company wanted to meet the needs of Middle Easterners who watch Western television when they travel but find nothing of interest on terrestrial television in their home countries.[8] Some years later, his successor, Samir Abdulhadi, confirmed Orbit's mission as 'a commercial business and not a public service enterprise'.[9] Saleh Kamel, the Saudi entrepreneur who launched Arab Radio and Television (ART) as a pay-TV network in the mid-1990s, initially differentiated ART's output from Orbit's by reference to the target audience. As the owner of the Dallah Al-Baraka conglomerate of businesses, he wanted to appeal to a conservative, 'silent majority' of Arabic speakers occupying a different market segment from the affluent English-speakers targeted by Orbit.[10] In contrast, early official statements about the rationale behind the al-Jazeera satellite channel seemed to conceive of the audience as a public. They expressed a concern to 'feel the people's pulse' and develop broadcasting institutions that would be consistent with parliamentary democracy.[11] Wadah Khanfar, al-Jazeera's director-general, said of the channel in 2006: 'Having a market-driven approach has never been a worry for us.' Referring to a long-standing advertisers' boycott of al-Jazeera, he added: 'Because of the boycott, we don't feel that advertising is a target. The target for us is professionalism. With [editorial] balance we gain a far larger audience.'¹²

Not everyone who watched the rise of Arab satellite television admired its consumerist dimension. Oliver Boyd-Barrett, having observed the planning stages of a new channel in Dubai, noted that the channel's distinctive 'Arab' image would not in itself signify a 'qualitatively different kind of relationship with its target audiences', since these would 'most likely continue to be positioned primarily as consumers'.¹³ What Boyd-Barrett overlooked was that treating Arab audiences as consumers marked a positive development compared with their treatment up to that point. A third approach – one that treated viewers as neither citizens nor consumers – had previously had a long history on Arab TV. Gamal Abdel-Nasser, in devoting precious resources to powerful radio transmitters for the pan-Arab station, Voice of the Arabs, in the 1960s, regarded the Arab 'masses' as a force to be mobilised by propaganda. Traces of an equivalent approach were still to be found in the 1990s and early 2000s, in references to the impact of satellite broadcasting on the so-called 'Arab street'. 'Arab street' was the expression routinely used about public responses to the al-Aqsa *intifada*, which erupted in September 2000, the aftermath of the September 2001 suicide attacks in the US and the Israeli military assault on Palestinian towns and refugee camps in March–April 2002. By using the expression 'Arab street' instead of 'Arab public opinion', commentators drew attention, sometimes inadvertently, to the huge imbalance in power between rulers and ordinary people. It was an imbalance mirrored in the unequal relationship between media owners and ordinary viewers, in which low-income viewers in non-oil countries without credible electoral systems lacked the power to make owners value them as either citizens or consumers. According to Samir Atallah, who compared the two concepts in 2002, the notion of the Arab street has an altogether different meaning from public opinion. Indeed, he said, it embodies Arab political impotence at all levels and thereby blocks the concept of public opinion from emerging. 'Public opinion', Atallah wrote, 'is formed in the media, in clubs, in open discussions, in free speech and involvement in the state's destiny. In the absence of these things – as is the case at present – the only resort is to "the street".'¹⁴

One overarching question, then, is how far the Arab television industry has moved in recent years to gauge, understand and respond to public opinion and public reception of the industry's output. A step in this direction is essential if broadcasters wish to claim that they are sensitive to audience preferences and needs. Measuring preferences involves not only quantitative but also qualitative methods. Statistics may serve advertisers' basic requirement to know the comparative popularity of certain channels, but discovering how audiences feel about particular programmes involves more than quantitative data. Qualitative feedback becomes even more important in markets dominated by a small number of powerful firms, because studies have shown that advertising-financed broadcasters in an oligopolistic industry have an incentive to compete by providing similar programmes rather than diverse ones, thereby reducing real choice.[15] These studies seem to confirm Hotelling's Law, which, applied to media, predicts that competitive media markets will tend to homogeneity, in the sense that broadcasters will produce multiple programmes that all conform to majority preconceptions.[16] Thus several similar mainstream programmes will each get a higher share of a large audience than a minority interest programme will get, even if it attracts all the audience with that minority interest. It can be seen from this that a competitive market, where ratings matter to both broadcasters and advertisers, may not result in the level of diversity or dynamism that might be expected in a situation where the viewer can accurately be described as king.

For Arab broadcasters to prove that they are really influenced by audiences, rather than the reverse, they would need to show a serious interest in commissioning polls and calculating programme ratings and in using the latest technology to maximise feedback from viewers. Given the tight links between Arab broadcasting companies and the region's governments or ruling elites, plus a long history of resistance to opinion polling, the showing of such an interest would be significant not only in terms of the media business itself but from a wider political point of view.

Trends in audience research

According to veteran commentator Rami Khouri, writing in September 2005, the 'ability to conduct public opinion polling' in many Arab countries counts as 'one of the truly historic recent developments in the Arab world in the past decade or so'.[17] It is easy to substantiate Khouri's pronouncement.

Just three years earlier, the polling group Zogby International had published *What Arabs Think*, an unprecedented survey of Arab opinions in which 3,800 adults from eight Arab countries, including Saudi Arabia, were asked ninety-two questions face to face about their values, political concerns, mood and outlook, self-definition and view of the world. In the words of its author, the study posed questions that had 'not been asked before'.[18] At around the same time a team of Arab scholars, backed by the United Nations Development Programme and the Arab Fund for Economic and Social Development, had begun conducting surveys and opinions polls for the first editions of the *Arab Human Development Report (AHDR)*. A poll for the *AHDR 2004* asked questions about freedom of opinion, expression and association. Although blocked in Egypt, it went ahead in five Arab countries that together accounted for more than seventy million people, or nearly a quarter of the total Arab population.[19] In March 2005, shortly before King Abdullah took the throne in Saudi Arabia, the Saudi government announced its readiness to survey Saudi public opinion. Abdullah's foreign affairs adviser, Adel Jubair, told a press conference in Washington that his government had commissioned a multimedia campaign against terrorism and planned to test its effectiveness through public opinion polls.[20] In May 2005, al-Arabiya announced that its weekly programme *Poll on Air* would be live for two hours every Friday at 2 PM Makka time. Nakhle El Hage, the channel's director of news and current affairs, described the move as a way to let Arab viewers 'set the agenda and tell us what they think is news ... As a station we need to establish a durable and trustful relationship with our viewers, which means being on the receiving end,' he said.[21]

In the specific field of media research, the Amman-based telecoms consultancy, Arab Advisors Group, expanded into television in August 2003, with a report on revenues earned from telephone voting during the first season of the *Super Star* singing contest on Lebanon's Future TV. Up to this point, the Pan-Arab Research Centre (PARC) and Ipsos-Stat had kept the market for television research to themselves. For some time afterwards they could still point dismissively to small survey samples in Arab Advisors Group research. In September 2004, for example, Arab Advisors Group produced a report for the Saudi pay-TV network Orbit based on a sample of just 134 respondents in four cities.[22] But this was soon to change. The Group's *Saudi Arabia Media Survey 2007* had face-to-face interviews with 674 respondents in six cities,[23] while its media surveys of Qatar and Kuwait in 2007 interviewed 530 and 605 respondents respectively.[24] Results on this scale hardly compare with the kind of electronic meter-based daily data collection that the Neilsen group operates in the US and elsewhere. Nevertheless, the Arab Advisors Group method of face-to-face interviews is noteworthy in comparison with the computer-assisted telephone interviewing (CATI) conducted by PARC and Ipsos-Stat, with which these companies say they generate around 130–150 interviews per day in Saudi Arabia.[25] Commenting on this method of data collection, with its monthly delivery of Saudi data to media clients (advertisers, TV stations, media agencies), three-monthly or annual intervals in data collation for other Gulf countries, and an absence of any external auditing of data, Ipsos-Stat's managing director Elie Aoun blamed 'politics' for causing people to doubt the accuracy of data.[26]

What this interviewing activity had yet to yield in any volume by early 2008 was systematic independent qualitative reports on audiences' opinions about programme content on Arab television. Meanwhile, attempts to introduce electronic metering in Gulf countries had repeatedly stalled. Although some agencies had tried to get meters up and running during 2007, Sam Barnett, MBC Group's chief operating officer, told an interviewer in January 2008: 'We anticipate people meters will go into Saudi Arabia in the next year or two years.'[27] In the absence of meters, there could be no consensus among users as to the credibility of audience statistics. In May 2004 a Zogby poll asked respondents in five Arab countries to name their first and second favourite news broadcaster. Zogby reported that al-Jazeera was 'by far the most-watched satellite channel for international news', with 'al-Arabiya most often the second choice of Middle East viewers'.[28] When al-Jazeera was again reported to be the most popular news channel in 2005, this time in six countries, Dubai TV, MBC, Saudi TV and Lebanon's Future TV all reacted negatively by criticising Zogby's methodology.[29] Al-Arabiya, part of the MBC group, went on the offensive with data supplied by Ipsos-Stat indicating that, between November 2005 and March 2006, it had not only turned the tables on al-Jazeera but opened up an eleven-point lead.[30]

As to whether advertisers themselves are convinced by such statistical findings or, more fundamentally, whether they even use statistical research at all to guide their advertising decisions, a leading executive in the business publicly confirmed in 2005 that decisions in television advertising are based on politics rather than numbers. The executive in question was Antoine Choueiri, head of the group that sells advertising airtime on behalf of so many apparently competing satellite channels that it is rumoured to control 70 per cent of the market.[31] Choueiri conceded to the Gulf Marketing Forum that he had decided not to renew his company's contract with al-Jazeera for 'political reasons'.[32] Links between politics and spending on advertising could also be deduced from PARC's annual report on 2006. With TV that year claiming a reported 43 per cent of regional ad-spend, against newspapers' 42 per cent,[33] governments were said to be among the biggest spenders on advertising in the television sector.[34]

At the same time, despite the boom in numbers of satellite channels, spending on television advertising remained very low in Arab countries compared with other parts of the world. PARC calculations for 2006 put spending on all forms of media advertising at an average of just US$102 per head across the six Gulf Cooperation Council (GCC) countries of Saudi Arabia, Kuwait, Bahrain, Oman, Qatar and the UAE in 2006. At US$375 and US$425 per head respectively, the equivalent figures for Europe and the US were around four times the GCCs'.[35] Many observers attributed the low level to advertisers' suspicions about a lack of competition between broadcasters, combined with a lack of confidence in television ratings. Tarek Ayntrazi, during a brief appointment with Future TV in 2005–6, was outspoken in his criticism of Choueiri's extensive control of the advertising market, blaming it for holding back growth. Describing the 'lack of transparent and reliable research data' as shameful, Ayntrazi said advertisers would opt for outlets other than television until the matter was resolved.[36] Paul Boulos, senior marketing manager for LBC, made similar points in late 2007. 'Unfortunately,' he complained, 'we are in a region where the blind leads the bland'. Has anyone asked the viewers what they want? Research is still at its infant stage: subjective and ego-driven. 'The media research budget in the region doesn't even account to (sic) 0.5 per cent of any advertising budget.'[37]

Is the Arab TV Viewer a King or a Pawn?

Trends in scheduling

If Boulos's estimate of low spending on audience data fails to accord with optimistic statements elsewhere in the industry about the Arab television viewer becoming king, it should be noted that there are similar contradictions between optimism and pessimism as regards trends in the quality of programmes. Booz Allen Hamilton declared in November 2007 that the proliferation of Arab channels, by stimulating greater competition, had 'begun to generate a greater diversity of programme offerings, including thematic channels, original productions and new programme formats'.[38] The company's researchers also found, however, that the competition process was privileging some aspects of programme production and not others. For example, producers of Egyptian television series, known for their dominant position in the industry over the past fifty years, were finding that the rising costs of star actors (now sought after by rich Gulf production companies and film directors) were eating into the amounts available to spend on good scripts.[39] Booz Allen Hamilton's conclusion was that large pan-Arab broadcasters seeking to retain their dominant position would need in future to 'provide their audience with high-quality programming to differentiate their offerings and stay relevant to the market', and that TV series would only capture larger shares in 'key advertising markets' in Saudi Arabia, Kuwait and the UAE if they were relevant to Gulf audiences.[40]

Relevance and difference could be found in programmes on Arab television in the 2002–6 period during which satellite broadcasters such as MBC, Dubai TV, al-Jazeera and others, which had long existed as single channels, were transforming themselves into multi-channel networks.[41] But indicators of relevance were more often to be found in efforts to silence popular programmes than in formal feedback from viewers. The silencing of the drama serial *Tariq ila Kabul* (*Road to Kabul*) in 2004 was particularly obvious as it happened when screening was under way and eight out of thirty episodes had already been shown. Under pressure from associates of the Taliban on one side[42] and US President George Bush's re-election campaign (worried about anti-American sentiment) on the other,[43] Qatar TV, which had commissioned the serial, refused to release the remaining episodes to channels that had bought it. Saudi Arabia's satirical comedy, *Tash Ma Tash*, a Ramadan series that started on Saudi state TV in the mid-

1990s and was later carried by MBC in recognition of its large and loyal following, has seemed to provoke talk of banning almost every year. In 2006 some religious scholars took such offence at its light-hearted treatment of religious extremists[44] that two episodes were reportedly cancelled and King Abdullah personally called on the show's creators to tread more carefully on subjects like tribalism and religion.[45] If Ramadan drama serials never make it past the censor to start with, as is said to have happened in at least five cases in Egypt, Kuwait, Saudi Arabia and Syria in 2007,[46] viewers do not even get a chance to form an opinion about the programme, let alone convey that opinion to broadcasters. In the many cases where they see a series after cuts have been made, it seems self-evident that the censor, not the viewer, is king.

Thus the declared upturn in respect for audiences in 2007–8 still had to be weighed against government interventions, including censorship threats enshrined in the Arab information ministers' February 2008 satellite broadcasting charter. It also had to be weighed against actual programming decisions. Waleed al-Ibrahim, chairman of the MBC Group, said in February 2008 that his group's 'permanent objective' was to present programmes that 'rise to the expectations of the Arab audience that is now more discerning towards the quality of programmes.'[47] But for some observers the idea of audiences becoming 'more discerning' was simply another way of saying that audiences had grown unmistakably tired of repetition and duplication on Arab satellite channels after several years in which entertainment programming seemed to be dominated by Western imports and reality TV formats.

Imports and formats bring with them an expectation of success based on audience figures obtained in countries where they have been screened or adapted so far. In other words, assumptions based on foreign audience data can take the place of research on the part of the importer. MBC started to rely on programme formats devised in Western capitals in the late 1990s, moving on to demonstrate in 2000 how a UK-originated format like *Who Wants to Be a Millionaire*, which had recently delivered an urgently-needed ratings boost for the US network ABC, could achieve a similar result for an Arab channel. Future TV followed suit in 2002, opting for another UK format, *Pop Idol*, which had achieved spectacular results for both the UK's ITV and Fox TV in the US before being reincarnated as *Super Star* on

Lebanese television. But there was one important difference between the *Millionaire* and *Pop Idol* formats. Telephone voting in the latter provided broadcasters with instant data about the programme's following – data capable of convincing advertisers to allocate their budgets to a show that was clearly attracting millions of young people with enough disposable income to own and use mobile phones. This feature offered another reason for rapid uptake of formats across the Arab television industry.

The phenomenal rise of formats was accompanied by a big increase in imports of Hollywood films, US sitcoms like *Friends* and *Scrubs* and confessional-style talk shows, including *Oprah* and *Dr Phil*, to fill space on new channels launched by MBC and Dubai TV during the period 2002–4, notably MBC2, MBC4 and One TV. Reality and lifestyle formats, like *Joe Millionnaire* or *Supernanny*, were also imported ready-made in their American versions and screened, like other US shows, with Arabic subtitles. For Ramadan 2005, MBC even tried out *al-Shamshoon*, an Arabised version of Fox TV's animated series, *The Simpsons*. This involved more than simply changing the characters' names (Homer to Omar, Bart to Badr and so on) along with their habits and settings and having Egyptian actors do their voices. Many commentators found the result unfunny and remained unconvinced that it was worth the effort. They wondered why local creativity could not be given more scope to develop indigenous ideas. Negative feedback on programmes surfaced ever more frequently through blogs. Yet leading Arab television conglomerates had their reasons for experimenting with shows like *al-Shamshoon* – aimed at young adults – alongside reality formats and ready-made imports. As Michel Costandi, MBC's business development director, explained in 2005, a major reason was segmentation, so as to help 'clients' (i.e. advertisers) reach their target market. MBC's plan, he said, was to attract children with a dedicated children's channel (MBC3) and 'be there with the graduation of our viewers' as they 'move onto MBC2, then MBC4 and then al-Arabiya'. 'Everything in the marketing world is about segmentation: to aim at an audience,' Costandi declared. 'Clients want a very specific market. It's a healthy thing to try and capture specialised, segmented audiences'.[48]

Conclusion

Statements by executives in the Arab television industry demonstrate how many of them adhere to a business model that does not give first or even second priority to viewers. First or second priority may go to advertisers, but it is not even clear exactly what level of importance leading broadcasters attach to non-government advertisers because they have been so lax for so long in gathering the kind of detailed ratings that profit-oriented advertisers need. The relative importance of advertisers is also somewhat obscured by evidence that government censors have the first say over what programmes are allowed on air and whether parts of programmes should be cut. Nevertheless it is clear from at least one channel executive's reference to audience segmentation that revenue requirements mean that advertisers' interests prevail in the planning of commercial network expansion and programme schedules. Segmentation regards viewers as consumers who can be divided into target markets, not as citizens who have collective interests as members of a public.

Finally, increasing talk of a need for differentiation between channels, heard from industry analysts and channel owners in 2007–8, seemed once again to bear out Hotelling's Law, in that the programming of leading players had clearly converged towards a point where many shows were similar to each other. It may be true that viewers were gaining more opportunities to express their opinions about television programming through blogs. But, given all the other factors considered here, the idea being expressed in some circles – that the viewer had become king – seemed rather far-fetched.

Notes

1. In this article, primary sources such as newspapers and online content are cited in endnotes but not the List of References.
2. This generalisation does not apply to the 2005 or 2006 Beirut Media Forum, at which early versions of parts of this article were presented.
3. The media research consultant in question is Houda Koussa, quoted by Nour Malas in 'Democratizing the Arab Screen', *Middle East Broadcasters Journal*, no. 16, January–February 2008.
4. For a more detailed investigation of rationales behind programming and the provision of audience data to advertisers see Naomi Sakr, *Arab Television Today*, London: I. B. Tauris, 2007.

5. See Martin Koschat and William Putsis, 'Who wants you when you're old and poor? Exploring the economics of media pricing', *The Journal of Media Economics*, no. 13 (4), p. 218.
6. Nicholas Garnham discusses issues raised by 'audience construction' in *Emancipation, the Media and Modernity*, Oxford: Oxford University Press, 2000, p. 117, based on work by Todd Gitlin in *Inside Prime Time*, 2nd ed., London: Routledge, 2000.
7. See, for example, Marc Raboy, 'Public service broadcasting in the context of globalisation' in Marc Raboy (ed), *Public Broadcasting for the 21st Century*, Luton: John Libbey Media, 1996, p. 5.
8. Interview with *Newsweek*, 6 June, 1994.
9. Interview with Samir Abdulhadi, *tbsjournal.com*, no. 9, Fall/Winter 2002.
10. Naomi Sakr, *Satellite Realms: Transnational Television, Globalisation and the Middle East*, London: I. B. Tauris, 2001, pp. 47 and 78.
11. Qatar's former information minister, Hamad Abdel-Aziz al-Kuwari, used the phrase 'feel the people's pulse' in relation to al-Jazeera in an interview with *al-Hayat*, published on 11 February 1996.
12. Author's interview Doha, 1 February, 2006.
13. Oliver Boyd-Barrett, 'Pan-Arab satellite television: the dialectics of identity', in H. Tumber (ed.), *Media Power, Professionals and Policies*, London: Routledge, 2000, pp. 330–31.
14. Samir Atallah, 'The Arab Street', *al-Nahar*, 18 September 2002. English translation in *Middle East Economic Survey*, 30 September 2003, pp. C2–3.
15. Colin Hoskins, Stuart McFadyen and Adam Finn, *Media Economics*, London: Sage, 2004, p. 209.
16. Jan van Cuilenburg, 'On competition, access and diversity in media, old and new', *New Media & Society*, no. 1/2, 1999, p. 196.
17. Rami G. Khouri, 'Arab democrats are denied the democracy they crave', *The Daily Star*, 21 September, 2005.
18. James J. Zogby, *What Arabs Think: Values, Beliefs and Concerns*, Utica NY and Washington: Zogby International/The Arab Thought Foundation, September 2002, pp. iii and 99.
19. UNDP Regional Bureau for Arab States, *Arab Human Development Report 2004: Towards Freedom in the Arab World*, New York: UNDP Publications, 2004, p. 98.
20. Reuters, 'Opinion polls taken in Saudi Arabia', 7 March 2005.
21. Quoted in 'al-Arabiya launches Poll on Air program', [http://business.maktoob.com/News-20050516144835-Al_Arabiya_launches_Poll_on_Air_program.aspx], 16 May 2005, (Accessed 18 February 2008).

22. Arab Advisors Group, *Saudi Sat TV and Radio Survey*, Amman: September 2004, p. 1.
23. Arab Advisors Group press release, 28 January 2007.
24. Arab Advisors Group press releases, 12 July 2007 and 7 November 2007.
25. Tim Addington, 'Dirty data', *Campaign Middle East*, 29 May, 2005.
26. Ibid.
27. Emirates Business 24/7, 'Most regional TV channels irrelevant', 29 January 2008. [www.business24-7.ae/cs/article_show_mainh1_story.aspx?HeadlineID=1594 (Accessed 2 March 2008)].
28. [www.bsos.umd.edu/sadat (Accessed 28 April 2006)].
29. *Asharq al-Awsat*, 'Leading Arab television stations reject Zogby report', 2 January 2006. Online at [asharqalawsat.com/English/article.asp?artid=id3285 (Accessed 25 June 2006)].
30. Alvin Snyder, 'Viewpoint: Al Arabiya as new favorite satellite TV', *Middle East Times*, 18 April 2006.
31. Without mentioning any name, PARC's 2007 report speaks of 'hushed up grouses' that 'one consolidated media group' has stacked up to 70 per cent of the market. PARC, *Harvest Y2007*, Dubai: Pan-Arab Research Centre, p. 11.
32. Tim Addington, 'Choueiri: "I am not the godfather"', *Campaign Middle East*, 13 November 2005.
33. PARC: *Harvest Y2007*, pp. 11–12.
34. PARC: *Harvest Y2007*, p. 22.
35. PARC: *Harvest Y2007*, p. 7.
36. Richard Abbott, 'The house that sales built', *Campaign Middle East*, 23 October 2005.
37. Paul Boulos, 'What revolution?', *Middle East Broadcasters Journal*, no. 15, November–December 2007 p. 36.
38. Booz Allen Hamilton press release, 13 November, 2007.
39. Ibid.
40. Chahine et al, *Trends in Middle Eastern Arabic Television Series Production*, Dubai: Booz Allen Hamilton, November 2007.
41. For background on the expansion process, see Naomi Sakr: *Arab Television Today*, pp. 168–188.
42. Peter Feuilherade, 'Media jihad', *Middle East International* 737, 5 November 2004, p. 15.
43. Charles Levinson, 'What's on during Ramadan? Antiterror TV', *Christian Science Monitor*, 3 November 2005.
44. The extremists were portrayed attending a Terrorism Academy, inspired by the Star Academy of LBC's weekly reality TV singing show. See Huda al-Saleh, 'Speak no evil: Muslim experts sound off over the ongoing "Tash"

controversy', *Asharq al-Awsat*, English edition, 10 October 2006, [http://www.asharqalawsat.com/english/news.asp?section=3&id=6659].
45. According to the Saudi-owned UK-based website Elaph on 17 October 2006.
46. Marlin Dick, 'True to form, Ramadan soaps stir controversy', [http://www.menassat.com/?q=en/news-articles/1921-true-form-ramadan-soaps-stir-controversy&page=0,1 11 February, 2008 (Accessed 29 February 2008)].
47. Salman Dossari, 'Q&A with MBC Chairman Al Waleed al Ibrahim', *Asharq Alawsat*, English edition 8 February 2008, [www.asharqalawsat.com/english/news.asp?artid=id11713].
48. Quoted by Gazbeya El-Hamamsy, 'Carving up the dish', *Business Today Egypt*, September 2005.

Bibliography

Arab Advisors Group, *Saudi Sat TV and Radio Survey*, Amman: September 2004.

Boulos, Paul, 'What revolution?', *Middle East Broadcasters Journal*, no. 15, November–December 2007.

Boyd-Barrett, Oliver, 'Pan-Arab satellite television: the dialectics of identity', in H. Tumber (ed.), *Media Power, Professionals and Policies*, London: Routledge, 2000, pp. 314–331.

Chahine, Gabriel, El Sharkawy, Ahmed and Mahmoud, Haitham, *Trends in Middle Eastern Arabic Television Series Production*, Dubai: Booz Allen Hamilton, 2007.

Garnham, Nicholas, *Emancipation, the Media and Modernity*, Oxford: Oxford University Press, 2000.

Gitlin, Todd, *Inside Prime Time*, 2nd ed., London: Routledge, 2000.

Hoskins, Colin, McFadyen, Stuart and Finn, Adam, *Media Economics*, London: Sage, 2004.

Koschat, Martin and Putsis, William, 'Who wants you when you're old and poor? Exploring the economics of media pricing', *The Journal of Media Economics,* no. 13 (4), 2001, pp. 215–232.

Malas, Nour, 'Democratizing the Arab Screen', *Middle East Broadcasters Journal*, no. 16, January–February 2008.

PARC, *Harvest Y2007*, Dubai: Pan-Arab Research Centre, 2007.

Raboy, Marc, 'Public service broadcasting in the context of globalisation' in Raboy, Marc ed. *Public Broadcasting for the 21st Century*, Luton: John Libbey, 1996, pp. 1–19.

Sakr, Naomi, *Satellite Realms: Transnational Television, Globalisation and the Middle East*, London: I. B. Tauris, 2001.

Sakr, Naomi, *Arab Television Today*, London: I. B. Tauris, 2007.

UNDP Regional Bureau for Arab States, *Arab Human Development Report 2004: Towards Freedom in the Arab World*, New York: UNDP Publications, 2004, p. 98.

Van Cuilenburg, Jan, 'On competition, access and diversity in media, old and new', *New Media & Society* 1/2, 1999, pp. 183–207.

Zogby, James J. Zogby, *What Arabs Think: Values, Beliefs and Concerns*, Utica NY and Washington: Zogby International/The Arab Thought Foundation, 2002.

JIM QUILTY

The Business of News: One Writer's Impressions of Two Middle East News Publications in 2005

Although *The Daily Star* – Lebanon's English-language daily newspaper – and *Middle East International* – a twice-monthly London-based specialist news magazine – are quite different from one another, both have embodied that mixture of idealism and pragmatism that characterises the business of journalism. Both have provided a news window onto the region for a (by most accounts) appreciative audience – much of which is located overseas, and which may have limited understanding of Arabic, Turkish and Farsi. To do so, each publication has had to make its own compromises, to strike a balance between survival and integrity.

The Daily Star and *Middle East International* are examined together because, as the writer has seven years experience with both publications, it seems their different capacities and choices provide a more comprehensive picture of the challenges that faced English-language journalism in this region in 2005. As the author has not worked in an editorial capacity at either publication, this paper should not be construed as speaking 'for' them in any way. The paper is less the matter of 'research' than anecdote and recollection. This is no airing-of-dirty-underwear exposé of incompetence and graft: all this information is a matter of public record.

The Daily Star

Since 2000, *The Daily Star* has been packaged on the newsstands with the *International Herald-Tribune* – an American newspaper that until a couple of years ago was jointly owned by the *New York Times* and the *Washington Post*. The *Times* later bought out the *Post*'s share. As *The Daily Star* is essentially a supplement of the *IHT* these days, the casual observer might assume that the relationship between the two newspapers is more intimate than it actually is. In fact *The Daily Star* has had an organic relationship with an international newspaper of note, but it wasn't the *IHT*.

The Daily Star was founded in 1952 by Lebanese journalist Kamel Mrowa. Locals will know him as the man who founded *al-Hayat*, the London-based Arabic-language newspaper that still enjoys an international readership. Mrowa was gunned down in 1966. Legend at the newspaper would have it the *Star*'s founder was killed by a supporter of then-Egyptian president Gamal Abdel Nasser. The assassin was never apprehended, a scenario this country has witnessed all too often, before and since.

Anecdotes from old hands at the paper suggest that, in the early days, there was some staff overlap between *The Daily Star* and *al-Hayat*. In the wake of Lebanon's civil war, that relationship changed when ownership of *al-Hayat* passed from Lebanese to Saudi hands.

For the decade after Mrowa's death the publication of the paper fell to his wife Salma al-Bissar, who carried on until the war forced its closure in 1976. The newspaper reopened as a daily from 1983–5, then ran as a weekly for another year before closing again. It returned to daily circulation in 1996. In its most-recent incarnation, *The Daily Star* has been in the hands of Mrowa's son Jamil Mroué, who holds the titles of publisher, editor-in-chief, and responsible director. After much negotiation with local authorities, Mroué closed a distribution agreement with the *International Herald-Tribune* in 2000. This agreement is based on a perceived community of interest.

For the *IHT* it is part of an ongoing effort to be the international newspaper of record in English, which has seen it make partnerships with little English-language newspapers from Europe to Japan. The arrangement allows the *IHT* to be published on site alongside its local partner, eliminating the postage-and-handling delays that make the hard copies of other (some would say superior) English-language dailies perpetually out of date for their expatriate audience. The market the *IHT* had in mind wasn't Lebanon alone, of course, but the region as a whole. Though it already had one partner in English-language *Haaretz*, the fact that this is an Israeli paper made it an inappropriate tool for penetrating the Arab market.

The *IHT*'s ambitions corresponded with those of *The Daily Star*, which itself needed to access a wider reading public (and with it a wider advertising market) than that afforded in Lebanon. The Lebanese newspaper's managers hoped to cash in on the prestige of the *IHT* brand, and were willing to pay for that privilege – too much, in the view of some who expressed scepticism of the deal's specifics.

The sealing of the arrangement roughly corresponded with *The Daily Star* going public, and at various points it has received cash infusions from different investors. This made it possible for the newspaper to divide in two – the *Lebanon Daily Star* and the *Regional Daily Star*. Both were packaged with the *IHT* and it was envisioned that, as other offices opened around the region, the *IHT-DS* package would include different local partners.

This arrangement survived for a couple of years until it was decided to collapse the two editions of *The Daily Star* into one another. The project didn't work (according to those employed to coax companies to buy ads within the new regime) because companies couldn't fathom buying advertising space in two newspapers of the same name: the Lebanon paper received ads, the regional paper did not. Since then, the *Star* has launched regional editions in the Arab Gulf and Egypt, so that it and the *IHT* now publish there as well.[1]

A brief examination like this one makes it impossible to discuss the travails the newspaper's management and staff, particularly the Lebanese staff, faced while navigating these waters of uncertainty, adjustment and experiment. A book could be written on the challenges *The Daily Star* faced in keeping itself afloat and its standards high. I will limit my observations to three, which are characteristic of any publication straddling two languages: the sales-readership disconnect; the expense of news gathering; headhunting.

When it was a strictly Lebanese newspaper, the *Star*'s struggle to obtain ads seems to have been particularly cut-throat – unfortunately this journalist's innocence of the details of Lebanon's advertising regime pre-empts a knowledgeable discussion of the matter.

Speaking anecdotally, one obvious challenge to selling companies' ads in *The Daily Star* has been that the relationship between sales and readership is less conventional than it is for Lebanon's Arabic-language newspapers like *al-Nahar* and *al-Safir*, which are deeply rooted in the socio-political landscape.

Without a reliable statistical breakdown of the demographic of *The Daily Star*'s readership, it seems that the vast majority of readers do so online, free of charge. Though many of these readers are probably Lebanese, most of them are expatriates. Most of the interested foreigners who read it – who seem to range from scholars of the Middle East to the well-meaning ignorant – also read it online.

At the time of writing, there are precious few, indeed no, stories about newspapers that have found a way to make their websites profitable while maintaining hits and reader feedback. In this *The Daily Star* is no different.[2]

Another challenge facing the *Star* is the sheer labour and expense of working in two languages. For reasons that have nothing to do with the quality of local talent, the number of journalists who can gather the news in Arabic and quickly generate clean copy in English are few indeed. Much of the copy the *Star* receives is in Arabic – often hand-written and faxed in from outside Beirut. That copy must then be translated into serviceable English, then copy-edited into something like journalistic English, a process that, if done too hastily, can easily imperil the intentions of all but the most straightforward news report.

In a perfect world those who perform the task of copy editing and subediting (the final on-the-page editing) should at least be native speakers of English; this puts further strains on the paper's resources because foreign employees require residency and work permits, neither of which are inexpensive.

These added layers of processing and expense absorb the resources of a paper that might otherwise be used putting more journalists in the field. This – not any editorial relationship with the *IHT* – is the reason *The Daily Star*, and the regional edition in particular, relies so heavily on wire copy, the same news agencies the *IHT* itself uses. This is also why the photographs adorning the *Star*'s front page are so often the same as the ones fronting the *IHT*: different editors who've received the same training for what makes a strong photo are drawing on the same pool of pictures and making similar choices. The problem of homogeneity of representation is, in this case, not one of editorial collusion, simply limited resources.

As mentioned above, intelligent journalists who can gather news in Arabic and write it in English are a rare commodity hereabouts, which leads to a third problem confronting a paper like *The Daily Star*: headhunting.

During this journalist's several-year tenure, a great deal of journalistic talent has passed through the newspaper. The most talented and ambitious do not stay because they invariably attract the attention of, and have been poached by, other media outlets ranging from *The New York Times* to Reuters and Bloomberg. More recently, some were attracted to offers from the Bush regime-sponsored Arabic-language television network al-Hurra. All were attracted to the resources of these institutions, if not their international profile.

Middle East International

For thirty-five years, *Middle East International* declared its mission to be 'to provide intelligent, authoritative and independent news and analysis on the Middle East.' The magazine was founded in 1971 by a number of retired British diplomats frustrated that the Western press tended to be predisposed to listening to one version of the Arab-Israeli conflict. *MEI* set out to present the 'Arab' version of the story.

When this author joined *MEI* – around the time that four Lebanese jurists were murdered in a Saida court by assailants who were never apprehended – the magazine's board of directors, 'The Knights', as the contributors know them, included Patrick Seale, the distinguished scholar on the Baath regime of Hafez al-Asad.

The magazine seemed to have limited its overheads by relying on contributions from journalists already on assignment for other media outlets. Many, like Jim Muir, were BBC journalists – though readers noted that *MEI* pieces by BBC journalists had more depth and analysis than the reports they filed with the World Service. Other contributors worked for local English-language publications or, like the stalwart Palestine contributor Graham Usher, were recognised as knowledgeable freelancers.

MEI's London office was manned skeletally by its long-time editor Steve

Sherman, an editorial assistant, and staffers paid to manage subscriptions and the website, an innovation brought online during the magazine's final years of life.

The website followed a policy, common in those days, of making two or three stories accessible to casual browsers, while the rest could only be got via passwords doled out to subscribers. It appears print subscriptions could have been better managed – both because the magazines arrived in the post rather too late to be useful to researchers and because, based on testimonials from several 'subscribers', it wasn't uncommon to receive copies of the magazine months, sometimes years, after allowing subscriptions to lapse.

Over the years, *MEI* found innumerable champions in the diplomatic and academic communities. More than most Western readers, they needed to be kept abreast of developments in the region without being hostage to the criteria that conditioned mainstream national media coverage – national interests (American, British, what have you), on one hand, and Israel on the other.

Over the years, *MEI* outgrew its original mission statement to simply 'tell the other side of the story'. Without ever abrogating its responsibility to tell the Palestine story in a balanced manner, it testified to the fact that there are other stories in this region besides the one in Palestine. It covered stories in the region from Mauritania to Iran (later the Caucasus, Afghanistan and Pakistan), indifferent to whether or not the stories conformed to mainstream media criteria. In doing so, it provided its readers with the tools to make their own analyses of the region's events, long before the coup or assassination or putative al-Qa'ida connection made it worthy of notice. It should be of no surprise, then, that in issue 760 Dennis Walters, the chairman of the *MEI* board of governors, announced that 'the issue of 28 October [2005 would] be the last.'

It seems a little counter-intuitive that such a publication should go out of business when the Middle East and related issues so dominated the mainstream press. The reasons Walters gave were informative on several levels:

> MEI has always run at a loss and the gap between income from sales and the cost of publication has been bridged by benefactions from

our supporters. In the present miserable situation in the Middle East those supporters have, understandably, more urgent calls on their generosity.

Moreover, our circulation has for some time been falling. This decline may seem strange at a time when the Middle East is front-page news every day. We are advised by experts that the reason is that news and comment are now available free and immediately on the Internet. All periodicals (and some newspapers) are facing the same problem and it is likely to get worse. In consequence raising money, even if it were possible to attract suitably large sums, would no longer help.

As Sherman relates it, when he came on board in 1987 the magazine was living on donations from various sources in the Arab world. The Saudis contributed £35,000 a year while another Gulf Emirate contributed £10,000 a year. The publishers of Sharq al-Awsat contributed several thousand pounds a year in return for a seat on the board of directors and sole, exclusive advertising rights – which came to nothing because *MEI* never ran ads.

In the 1990s these contributions dried up when the magazine raised something like £250,000 from some Palestinian donors. The Saudis, meanwhile, increased their donation to £50,000 a year at about the same time that the other Gulf donor stopped contributing to the pot.

The 2005 financial crisis arose from the donors' not donating. *MEI*'s news service likely failed because – in an effort to remain autonomous from market forces – it relied upon arbitrary financial contributions. These resources not only proved unreliable, they forced the magazine to live a hand-to-mouth existence that made outreach campaigns like advertising too expensive.

Sherman told me that the following week he would hear back from his last hope for a short-term bailout. All who profess to care about informed opinion about this region should be saddened that this eventuality was allowed to come to pass.

Notes

1. In the years intervening between the time of writing and the publication of this collection, *The Daily Star* ended its distribution arrangement with the *IHT*, the local *Daily Star* franchises, which were effectively independent, changing their names.

   The arrangement hadn't been without its problems, in any case. When the *IHT* ran ads terming Hizbullah a terrorist organisation, for instance, or expressing support for the state of Israel (with which Lebanon was then at war), the task of filling the space with less-provocative ads proved too complex. Instead, that issue of the *IHT* was simply not printed, putting the Lebanese newspaper in breach of contract.

   The distribution arrangement was finally put to death after 2006. It was impossible for *The Daily Star* to print the *IHT* for the duration of the 34-Day War of July–August 2006 (it seems newsprint supplies were low – *The Daily Star* itself being reduced to four pages by the end of the conflict). Afterwards, it proved impossible to return to the previous regime, due to low revenues attributed to the collapse of the local ad market.

2. In 2006, *The Daily Star*'s IT department staff undertook to make its website profitable by archiving all published stories after a few days and making the archive available to subscribers only. Without discussing the various complaints readers expressed about the newspaper's website, this measure corresponded to a drastic reduction in the number of hits the site received, arguably reducing its profitability.

TRISTAN MATTELART

Transnational Media and Authoritarian National Public Spheres

Transnational media have often been viewed as a danger to national identities or sovereignties. They are rarely considered from the perspective of the strategic voice that they might provide for countries in which the regime, jealous of its ideological monopoly, strives to control the local means of expression.

I seek to emphasise the importance of studying the transnational media's ability to circumvent domestic censorship policies. My aim is to show how these media can, from the outside, nourish national public spheres which, in authoritarian contexts, are under surveillance. My focus is on the way in which international radio and television have been able to feed, in different geopolitical realities, the resistance strategies of dissidents who are forbidden to speak in the official public sphere or to contribute, in these same realities, to the tactics that the population uses to access other representations of the world.[1]

National censorship threatened by audiovisual media

The literature on media internationalisation has until recently largely ignored the potential strategic functions of transnational means of communication in authoritarian regimes. Pioneering studies in the field of critical political economy were the first to demonstrate the existence of a transnational media system and to highlight the logics of domination that organise it.[2] However, these studies tended to present the ubiquity of Western means of communication as a menace for the cultural identity of the nations confronted with it.

In the context of the 1970s, the defence of national identity, of national culture and of cultural autonomy of the countries of the Third World featured prominently on the scientific agenda of critical thought. In the famous book they edited in 1979, *National Sovereignty and International Communication*, Kaarle Nordenstreng and Herbert I. Schiller go so far as to suggest that the concept of national sovereignty is an essential tool for the study of international communication processes.[3]

As important as they have been in raising awareness of the challenges posed by media internationalisation, the concepts of national identity and national sovereignty are not without ambiguities. They have served to justify the defence of the cultural industries of the South subjected to the logic of unequal exchange. However, they have also been used by regimes willing to impose their monopoly over expression to legitimise the censorship of transnational audiovisual flows.

Furthermore, one of the problems of this analytical framework is that it allows us to think of transnational audiovisual media only in terms of the economic, political and cultural dependency that they create for the countries that receive them. Usually ignored is the capacity of these media to evade the barriers that censors impose and, in this manner, to satisfy certain democratic demands.

Scarce, in this framework, are critical researchers interested in international radio on short- and medium-wave frequencies, even though these represent a major and rather long-term player in the transnationalisation of news and entertainment.

The rare literature dedicated to international radio, often written by former employees of these radio stations, underlines the strategic role they play in authoritarian regimes. Donald R. Browne explains in his 1982 reference book on international broadcasting that if 'international radio broadcasting has been and remains a significant form of international communication', it is because it can 'work in the [...] "direct-from-broadcaster-to-audiences" manner', ignoring the filter of national borders – something that international television was not able to do before the second half of the 1980s.[4]

International broadcasters are therefore disturbing influences for a number of regimes that wish to preserve their ideological monopoly. These radio stations limit the autonomy of authoritarian governments by providing an alternative voice, and by feeding their listeners with 'news and opinions of a kind that the domestic media fail to provide,' Julian Hale observes in *Radio Power*. He even writes that without the international broadcasters 'the politicians [of these countries] would then be able to govern in their own way, rather than be forced to react to the pressures of public opinion.'[5]

The study of international broadcasting thus suggests a new manner of comprehending the functions of the transnational audiovisual media, by looking at their capacity to circumvent national censorship policies.

Western broadcasters are not the only ones to help the listener overcome national restrictions on information. Douglas Boyd shows in his book *Broadcasting in the Arab World* that Egyptian radio (which, under the regime of Gamal Abdel Nasser, relied on an international service to disseminate its pan-Arab ideology, the Voice of the Arabs, that enjoyed a large popularity in the Arab world) was, during the 1980s, still in a position to 'compete with the main international broadcasters in the region': the BBC, the Voice of America or RMC Moyen-Orient.[6]

Browne remarks for his part that at the beginning of the 1980s, the Cuban station Radio Habana was one of the most popular Spanish-speaking international radio stations in Latin America. This was due to the publicity it created for 'the activities of the opposition in diverse Latin American countries' governed by military regimes[7] – activities that went unnoticed in the local media ... and in the programmes of the Voice of America.

Indeed, international broadcasters are in the majority financed by public funds and are, for this reason, more or less subject to the diplomatic aims of those that they depend on; a fact that considerably constrains their capacity to fill in the information gaps in certain foreign countries. If the Voice of America denounces human rights violations in diverse states, it does this 'based on political criteria and perceived national interests', accuses Laurien Alexandre in 1988, in one of the few critical works dedicated to international radio broadcasting. The Voice of America never failed to highlight the 'Soviet abuses of civil liberties' in the course of its history but did not give 'the same attention to the brutal torture and killing of political prisoners in Chile (under Pinochet), the Philippines (under Marcos), Iran (under the Shah), or South Africa (under apartheid), to name a few; all these [human rights] violations [were committed] with the tacit support of the US government.'[8]

Although rare, there are a few critical studies about the contribution of international radio to the circumvention of local censorship policies. Spain under Franco served as a 'laboratory' in which to study the way in which international radio stations (but not the Voice of America!) can assist the 'clandestine communication' of the opposition in a dictatorial regime. Lluís Bassets and Enric Bastardes, in particular, highlighted the importance of 'the role played by the Spanish programmes on stations from socialist countries – notably Radio Prague and Radio Moscow – [as well as] the Spanish news from Radio Paris and the BBC, with a much more rigorous and truthful treatment of information.' Alongside the clandestine communist station Radio España Independiente, they played an important role in animating the underground resistance against Franco.[9] Bassets considers – anticipating research perspectives developed in the beginning of the 1990s – that by disseminating in Spain an opposition culture that was often preserved in exile, the radio stations broadcasting from abroad participated in feeding the 'clandestine and illegal sphere' of the opponents to Francoism. The role of this clandestine sphere was, he writes, 'to occupy the space of the public sphere' of the authoritarian state in order to publicise opinions that were circulating covertly.[10]

As much as they have forsaken the study of the international radio broadcasting, critical researchers have neglected the clandestine transnational television flows and prioritised the official ones. To study these televised 'pirate' flows, one needs to move away from the call to defend national sovereignty and to explore, instead, the way they contribute to the undermining of state policies of entertainment control.

In 1982, Boyd showed how, during the 1970s in the Gulf countries, video recorders enriched a limited official supply of televised programming. The 'videocassette business' in Saudi Arabia, halfway between legality and illegality, helped to circumvent 'the government's policy of controlling the kind of visual material shown in the kingdom', through its rich choice of American, British and also Egyptian films or broadcast programmes.[11]

Another book that underlines the importance of clandestine television flows was published in 1989 by Boyd, Joseph D. Straubhaar and John A. Lent. It studies the spread of *Videocassette Recorders in the Third World*. The importance of video tape recorders, write the authors, is due to the possibility they offer the audiences 'to circumvent systemic or governmental controls on [television] content'... to the great benefit of American content producers. The authors' research shows indeed that the programmes circulating worldwide on video are predominantly Hollywood productions.[12]

Radios and dissident strategies

With the growing strength of discourse about globalisation, by the end of the 1980s and the beginning of the 1990s, the intellectual environment became more and more conducive to a critique of the concepts of national identity and sovereignty. Nordenstreng and Schiller noted this change in the work they directed in 1993, revealingly entitled *Beyond National Sovereignty*.[13] In his conclusion, Nordenstreng recognises that, while in the 1970s the concept of national sovereignty 'was still taken more or less for granted', it now appears paradoxical. It serves '*both* as a shield protecting people's authentic interests against foreign domination *and* as an instrument of people's repression by national elites.'[14]

In this context other studies, appearing in the 1990s, also break with the analytical framework that focuses on the dangers that audiovisual internationalisation presents for national sovereignty. Focusing on the functions of the communication means in the Iranian revolution of 1978–9, on the disintegration of the so-called Eastern regimes in 1989, or on the democratisation of certain countries in the Third World, these works emphasise the role transnational media can play in reinforcing 'opposition spaces' and in the 'political mobilisation' against repressive states.[15]

Jaques Semelin, in his book *La liberté au bout des ondes,* examines more specifically the 'determining role' that Western stations had 'in the development of the resistance in the Soviet bloc'. Western broadcasters were players of the highest importance in this resistance because they represented privileged means of access to other worldviews, but also because they offered a sort of forum to the different opposition movements – not without struggles for influence – to address their own citizens. Opposition movements in different Eastern European countries were indeed more or less isolated and did not possess a network of communication to fully cover the national territory. Under these conditions, the dissidents, by bringing their news and publications to the West, were able to use Western radio stations to be heard in the four corners of their own country, or at least this was the objective. Semelin invokes in this respect 'a new way to communicate in the East via the West, an East-West-East communication'. The focus here is largely on the utility of the 'practices of interference of the media' of the West in the East.[16]

My own work, *Le cheval de Troie audiovisuel*, published in 1995, also paid attention to the way Western radio stations accompanied the strategies of dissidents' resistance in Eastern European countries. The book shows how, faced with the 'occupation of the public sphere by the state', the dissidents of Central Europe strove from the 1970s to create a sort of 'parallel public sphere' where criticism of state power could be voiced. They were able to do so through the more or less dynamic medium of parallel cultural and intellectual activities. The objective was to nurse the renaissance of 'civil society' and to force the regime towards democratisation. By giving a larger echo to their parallel publications, Western radio naturally constituted a partner of choice for the dissidents.[17]

In the book he devoted to the role of the media in the political transitions in Russia, Poland and Hungary between 1980 and 1995, John D. H. Downing also notes the decisive relays offered by the BBC, the Deutsche Welle, Radio Liberty, Radio Free Europe and the Voice of America for the 'alternative communications' of the opposition and as such for the animation of 'alternative public spheres' that they constitute.[18] 'Yes, of course, [these radio stations] were funded and designed to further the global interests of US and Western corporate powers, but to recognise only that dimension of their operation is to be singularly myopic.'[19]

These works are far from interpreting the Western radio broadcasters as a menace to the national identity of the different countries east of the Iron Curtain. Rather, they describe the role the broadcasters play in the preservation and dissemination of a dissident culture that, even though clandestine, represents no less an essential component of the identity of these nations.

These mechanisms that came to light in the context of the West-East relations are also found in certain studies on the way international radio stations contributed to the wave of democratisation which touched various countries of French-speaking Africa at the beginning of the 1990s. In his study of the influence of international radio broadcasters on the Beninese political crisis, Theophile Vittin shows how Radio France Internationale, Africa No 1, the BBC, the Voice of America or the Deutsche Welle 'echoed the declarations and positions of opposition and dissenting groups'. Better, he notes how the large public exposure that these radio stations gave to the fall of the Berlin wall contributed to 'the politicisation of the demands' of the Beninese opposition, and accelerated the fall of the regime of Mathieu Kérékou.[20]

Another situation in which to study the strategic role of the transnational media is the Iranian revolution of 1978–1979. In their reference book on the subject, *Small Media, Big Revolution*, Annabelle Sreberny and Ali Mohammadi analyse, firstly, the way audiocassettes, photocopied leaflets and traditional means of communication served as 'vehicles for a speech of opposition' to the Shah's regime. They show how these media took part in building a dissident political 'space' outside the state-controlled public sphere, and succeeded in mobilising a mass movement. However, the authors also underline the important role played by international radio that nurtured this opposition sphere. International broadcasters, especially the BBC, gave opponents of the regime – those in Iran as well as those in exile – the opportunity to address the Iranian population. As such, they offered a national means of communication to the 'political players within the country'. Thus, the programmes of the BBC became 'a necessary intermediate stage in what [was] really domestic political communication.'[21]

Transnational television and cultural resistance

In the preceding paragraphs, priority was given to the study of the participation of international radio stations to 'processes of active resistance' or, more precisely, to 'organised manifestations of resistance'. The work of James C. Scott invites us nevertheless to pay attention to non-declared modes of resistance, what he calls 'everyday forms of resistance', this 'wide variety of low-profile forms of resistance that dare not speak in their own name.' This American political scientist scrutinises ordinary practices of ideological insubordination that develop 'offstage', in social spaces shielded from the surveillance of the dominant power. These practices of ideological insubordination give substance to what he calls 'the infrapolitics of the powerless'. Rather than developing openly, in more or less formal political organisations, these 'infrapolitics of the powerless' take shape in 'informal networks of kin, neighbours, friends and community.'[22]

If the transnational audiovisual flows are in a position to support strategies of resistance of the opposition in repressive regimes, they can equally contribute to the 'infrapolitics of the powerless', to these forms of ordinary resistance, more covert, but no less important because they constitute the fertile ground on which more organised and institutionalised acts of opposition could develop.

Thus, *Le cheval de Troie audiovisuel* shows how international radio stations east of the Iron curtain, fed the tactics of 'auto-information' performed by the populations in order to compensate for the insufficiencies of the official communication apparatus. In so doing, these radio stations nurtured the formation of informal and parallel communication circuits. These evolved in social circles of proximity that allowed and extended, in the private sphere, the reporting of news that was banned from the public sphere.[23]

The transnational media do not only play a strategic role through their news programmes, but also through their entertainment programmes, which carry representations of social realities, even if these are often truncated. *Le cheval de Troie audiovisuel* describes the manner in which entertainment programmes of Western radio stations, of Western European cross-border television and – from the 1980s on – of video representing an attractive perspective of Western ways of life were spread in the East. In doing so, they participated in the dissemination of visions of the world in contradiction of those officially propagated in the societies of the East.[24] While I abstain from using the term 'resistance' to qualify the possible subversive uses of transnational audiovisual flows, other studies do not hesitate to take recourse to this terminology.

'Cultural resistance' is the term Sreberny and Mohammadi use to describe the tactics implemented by the Iranians at the beginning of the 1990s to access a 'global popular culture stemming mainly from the West, [...] distinct from and mainly in opposition to the "official culture of Islam" ... For many Iranians, in the absence of democratic rights in the Islamic Republic', the consumption of prohibited videocassettes is a mode of 'protest' against the power.[25]

The multiplication of satellite channels since the second half of the 1980s[26] opened a new field of study to those interested in the way transnational audiovisual flows contribute as a resource to the practices of circumventing state monopolies on news and entertainment. Riadh Ferjani describes the 'ruses' of Tunisian viewers, benefiting from videotapes and satellite antennas since the beginning of the 1980s and 1990s respectively, to escape monotonous national television programming. 'The usage of the tape-recorder and of the antennas to receive foreign stations,' he concludes, 'has breached the monopoly of audiovisual programming. They finished by structuring themselves into real networks of parallel communication, in large parts eluding the state powers.'[27]

Lotfi Madani studies the 'diverse strategies of searching for other images', such as the satellite dish that Algerians used in the 1990s to access entertainment and fiction programmes they liked ('Brazilian or Mexican *telenovela*, American or French sitcoms, Egyptian police series or soap operas'), as well as the news programmes offered by French and soon also by Arabic-speaking channels.[28]

In her book on transnational television in the Middle East, Naomi Sakr shows that in this region, it is through the contact with CNN at the beginning of the 1990s that viewers realised their dissatisfaction with the news on their own national channels. She explains to what extent 'CNN's live international broadcasts from Baghdad during the 1990–1991 Gulf crisis' revealed 'the stark contrast between instant, live television with minimum commentary' represented by the Atlanta channel and 'the stale, turgid and censored coverage available on local Arab stations'.[29]

These studies, in particular those by Ferjani and Madani, also demonstrate to what extent the relationship of Arab viewers to Western images was conflicted during the 1990s. The bias in the news on French networks about the events in the Maghreb or about immigrants living in France were such as to draw mistrust, suspicion and even anger from their Tunisian and Algerian viewers. These sentiments of mistrust regarding French reporting explain why the public in the Maghreb, wishing to escape the weightiness of their national news, have turned more and more to pan-Arabic channels since the mid-1990s.[30]

Because the geographies of the audiovisual flows have considerably changed since the 1970s and the first analyses on communication and cultural domination by Schiller, one must henceforth account for new media powers, including those which emerged in the countries of the South.

The major role that al-Jazeera plays as a supplier of news in the Arab world and beyond is revealing in this respect. Having been dethroned by the Qatari network and other pan-Arab stations inspired by it, Western satellite channels have not been the central players circumventing the state monopolies in Middle Eastern news since the second half of the 1990s.

Al-Jazeera owes its success in the Arab world largely to its capacity to bring news that is live, contradictory and ignores governmental censorship (albeit complying with the Qatari one!). 'Al-Jazeera has broken the taboo of the incontestable and uncontested, singular truth of the Arab regimes,' explains Olfa Lamloum. '[It has], for the first time in the modern history of the region, rendered visible and audible Islamist, nationalist and feminist oppositions.'[31]

The popularity of the Qatari network in the Arab world also comes from its capacity to hold a discourse about the conflicts stirring the region that is different from the one given by the Western media. Al-Jazeera, argues Lamloum, contests 'the dominant Western media's monopoly of sound and narrative about the Arab world and the world in general'. The Qatari station proposes 'a reading that is no less ideological than that of the Western media, but that reflects the views of the dominated peoples of the Middle East and thus finds itself better adapted to their specific demands.'[32]

Even though the attention of studies about the role of cross-border media in the Arab world has focused largely on issues related to television since the 1990s, this region shows how much the role of the radio, as elsewhere, remains important.[33] The launch of Radio Sawa in 2002 by the American government in an attempt to win the hearts and minds of the Arab population testifies to these dynamics.

If there could still be any doubt about it, this initiative – as well as the creation of al-Hurra in 2004, the inauguration of Arabic programmes on France 24 in 2007 and the inauguration of BBC Arabic TV in 2008 – shows that it is of capital importance, when considering the challenges represented by cross-border audiovisual media, to also focus on the diplomatic functions that can be attributed to them.

An 'emancipating' role?

This new way of thinking about the role that transnational audiovisual media can play – emphasising their participation in the circumvention of state monopolies of communication – is not without danger. When valuing the strategic functions that these media have in certain countries, there is a risk of underplaying the relevance of a critical approach toward them.

This is the case in an article by Sreberny, in which she plays down certain analyses of the political economy in order to highlight the capacity of satellite television to act in favour of democratisation in Middle Eastern countries. Rather than assuming, as critical scholarship tends to do, that 'commercialised media dynamics' render democratic debate more fragile, she suggests that 'we should be interested whether new political ideas can actually be provoked/promoted by the same dynamics ... Ironically, in a region where the dynamics of civil society have been somewhat weak, commercial media [regional as much as Western commercial satellite television] might bring more news, information and debate than many state systems have previously allowed.' In the same manner, instead of focusing on 'Western cultural domination' induced by the considerable presence of Anglo-Saxon programmes on national and transnational television in the region, one should be interested in the positive role that Western content can play. 'Television, especially with significant doses of foreign programming, displays the variety of lives in the world, including the far greater individualism, freedom and emancipation of women in Western societies (as well as in some Arab societies in comparison to others).'[34]

While it is necessary, as we have shown, to take into consideration the useful role that transnational audiovisual flows can play in authoritarian regimes, one must, nonetheless, not underestimate the problems that arise with the logics of domination and commercialisation that thoroughly structure the internationalisation of the media.

In this sense Edward S. Herman and Robert W. McChesney recognise the 'positive effects' of the global media and their ability to carry across borders some values 'such as individualism, scepticism of authority, and, to a degree, the rights of women and minorities' that can 'disturb authoritarian governments and repressive traditional customs'. Still, they condemn with great force the global process of commercialisation in which the media increasingly respond to consumer needs rather than to citizen needs.³⁵

It is all the more important to keep oneself from quickly attributing emancipating virtues to Western transnational media, since several studies show how very different authoritarian governments, taking note of the potential subversive effects of the programming of these media, have tried to employ them for social pacification – even if this attempt at domestication may turn against them.

The critical South African researchers Keyan and Ruth Tomaselli analysed how, from 1979 onwards, the apartheid regime, aware of the need to stimulate the emergence of a black bourgeoisie – in order to divide the black community and reign better – encouraged the South African Broadcasting Corporation to import American series starring black actors, such as *Archie Bunker's Place, Maude, Magnum, Starsky and Hutch* or *Today's FBI*. Obviously *Starsky and Hutch* could, in many ways, appear subversive in the apartheid country: Do these 'two smooth, hip, plainclothes policemen' not take orders from a black boss? Nevertheless, the discourse of the series, in the eyes of the Tomasellis, was largely in line with the objectives of Pretoria. Do Starsky and Hutch not have a mission to 'rid [their country] of criminals and people who threaten the established order'? And does their black boss not 'stand for the "successful" petit bourgeois black who has "made it"'?³⁶

Le cheval de Troie audiovisuel for its part brings to light the policies under which Western entertainment programmes were co-opted by diverse regimes east of the Iron Curtain in the 1970s and in the 1980s. Their goal? To provide sufficiently attractive imported televised content with the aim of 'divert[ing the population] from the public sphere, [...] to confine it to its private space', and to keep it far from any type of socio-political dispute.³⁷

Finally, Dominique Colomb highlights in his work *L'essor de la communication en Chine* the capacity of the Chinese state to utilise Western advertising and television know-how to its own benefit, to promote the illusion of 'commercial happiness', all the while keeping up one of the world's most repressive political systems. Christian Coméliau notes in his foreword that 'the merchant ideology is not only the successor of the [Chinese] state ideology, but has become its ally.'[38]

If research about the way transnational media participate in the circumvention of state monopolies on information and entertainment helps to better understand the variety of roles they can play in international relations, it requires also a certain critical vigilance. It is necessary to adopt a critical approach to research of the strategic functions of transnational media, as sketched above. This critical approach needs to be attentive to the reconfiguration of dominant strategies, to focus on the realities of an exchange that is still unequal and to look carefully at the logic of commercialisation that accompanies, in the North as much as in the South, the growing transnationalisation of the audiovisual.

Notes

1. For a more comprehensive approach see Tristan Mattelart, *La mondialisation des médias contre la censure. Tiers Monde et audiovisuel sans frontières.*
2. Herbert I. Schiller, *Mass Communications and American Empire* and *Communication and Cultural Domination.*
3. Kaarle Nordenstreng and Herbert I. Schiller, *National Sovereignty and International Communication.*
4. Donald R. Browne, *International Radio Broadcasting. The Limits of the Limitless Medium*: pp. II and 2.
5. Julian Hale, *Radio Power. Propaganda and International Broadcasting*, pp. 165 and 172.
6. Douglas Boyd, *Broadcasting in the Arab World. A Survey of Radio and Television*, p. 13.
7. Donald R. Browne, *International Radio Broadcasting. The Limits of the Limitless Medium*, p. 262.
8. Laurien Alexandre, *The Voice of America: From Detente to the Reagan Doctrine*, p. 28.
9. Lluís Bassets and Enric Bastardes, 'La prensa clandestina en Catalunya: Una reflexión metodológica', pp. 157–158.
10. Lluís Bassets, 'Clandestine Communications: Notes on the Press and Propaganda of the Anti-Franco Resistance, 1939–1975', pp. 199–200.
11. Douglas A. Boyd, *Broadcasting in the Arab World. A Survey of Radio and Television*, p. 143.

12. Douglas A. Boyd, Joseph D. Straubhaar and John A. Lent, *Videocassette Recorders in the Third World*, pp. 3–4, 14, 143.
13. Kaarle Nordenstreng and Herbert I. Schiller, *Beyond National Sovereignty: International Communication in the 1990s*.
14. Kaarle Nordenstreng and Herbert I. Schiller, *Beyond National Sovereignty: International Communication in the 1990s*, p. 461.
15. Annabelle Sreberny-Mohammadi, 'Introduction', p. 9.
16. Jacques Semelin, *La liberté au bout des ondes. Du coup de Prague à la chute du mur de Berlin*, pp. 12–21.
17. Tristan Mattelart, *Le cheval de Troie audiovisuel. Le rideau de fer à l'épreuve des radios et télévisions transfrontières*, pp. 57–87.
18. John D. Downing, *Internationalizing Media Theory. Transition, Power, Culture. Reflections on Media in Russia, Poland and Hungary, 1980–95*, pp. 74–78.
19. John D. Downing, *Radical Media. Rebellious Communication and Social Movements*, p. 360.
20. ThéophileVittin, 'Les médias externes comme facteurs de renforcement de l'opposition interne au Bénin (1987–1992)', p. 148.
21. Annabelle Sreberny-Mohammadi and Ali Mohammadi, *Small Media, Big Revolution. Communication, Culture and the Iranian Revolution*, pp. 30 and 133.
22. James C. Scott, *Domination and the Arts of Resistance. Hidden Transcripts*, pp. 4, 19 and 200.
23. Tristan Mattelart, *Le cheval de Troie audiovisuel. Le rideau de fer à l'épreuve des radios et télévisions transfrontières*, pp. 74–87.
24. Ibid., pp. 121–203, see also Tristan Mattelart, 'Transboundary flows of Western entertainment across the Iron Curtain'.
25. Annabelle Sreberny-Mohammadi and Ali Mohammadi, *Small Media, Big Revolution. Communication, Culture and the Iranian Revolution*, pp. 179–187.
26. Jean K. Chalaby, *Transnational Television Worldwide. Towards a New Media Order*.
27. Riadh Ferjani, 'Usages des nouvelles technologies de l'information et de la communication en Tunisie', pp. 33–34 ; see also Riadh Ferjani, 'Internationalisations du champ télévisuel en Tunisie'.
28. Lotfi Madani, 'L'antenne parabolique en Algérie, entre dominations et résistances', pp. 196, 203.
29. Naomi Sakr, *Satellite Realms. Transnational Television, Globalization and the Middle East*, pp. 10, 84–85.
30. Riadh Ferjani, 'Internationalisations du champ télévisuel en Tunisie' and Lotfi Madani, 'L'antenne parabolique en Algérie, entre dominations et résistances'.

31. Olfa Lamloum, *Al-Jazeera, miroir rebelle et ambigu du monde arabe*, p. 33.
32. Ibid., p. 35.
33. Olfa Lamloum, *La restructuration de l'espace radiophonique arabe. Palestine, Liban, Syrie, Jordanie et Égypte*.
34. Annabelle Sreberny-Mohammadi, 'The Media and Democratization in the Middle East: The Strange Case of Television', pp. 195–197.
35. Edward S. Herman and Robert W. Mc Chesney, *The Global Media. The New Missionaries of Corporate Capitalism*.
36. Keyan Tomaselli and Ruth Tomaselli, 'Between policy and practice in the SABC, 1970–1981', pp. 135–138.
37. Tristan Mattelart, *Le cheval de Troie audiovisuel. Le rideau de fer à l'épreuve des radios et télévisions transfrontières*, pp. 164–165.
38. Dominique Colomb, *L'essor de la communication en Chine. Publicité et télévision au service de l'économie socialiste de marché*, p. 13.

Bibliography

Alexandre, Laurien, *The Voice of America: From Detente to the Reagan Doctrine*, Norwood: Ablex, 1988.

Bassets, Lluís, 'Clandestine Communications: Notes on the Press and Propaganda of the Anti-Franco Resistance, 1939–1975' (first published in 1976), in Mattelart, Armand and Siegelaub, Seth, eds, *Communication and Class Struggle. Vol. 2. Liberation, Socialism*, New York and Bagnolet: International General-IMMRC, 1983, pp. 192–200.

Bassets, Lluís and Bastardes, Enric, 'La prensa clandestina en Catalunya: Una reflexión metodológica', in Vidal Beneyto, José, ed., *Alternativas populares a las comunicaciones de masa*, Madrid: Centro de investigaciones sociológicas, 1979, pp. 155–175.

Boyd, Douglas A., *Broadcasting in the Arab World. A Survey of Radio and Television*, Philadelphia: Temple University Press, 1982.

Boyd, Douglas A., Straubhaar, Joseph D. and Lent, John A., *Videocassette Recorders in the Third World*, White Plains: Longman, 1989.

Browne, Donald R., *International Radio Broadcasting. The Limits of the Limitless Medium*, New York: Praeger, 1982.

Chalaby, Jean K., ed., *Transnational Television Worldwide. Towards a New Media Order*, London: I. B. Tauris, 2005.

Colomb, Dominique, *L'essor de la communication en Chine. Publicité et télévision au service de l'économie socialiste de marché*, ('Foreword' by Coméliau, Christian), Paris: L'Harmattan, 1997.

Downing, John D. H., *Internationalizing Media Theory. Transition, Power, Cul-*

ture. *Reflections on Media in Russia, Poland and Hungary, 1980–95*, London: Sage, 1996.

Downing, John D. H. et al., *Radical Media. Rebellious Communication and Social Movements*, London: Sage, 2001.

Hale, Julian, *Radio Power. Propaganda and International Broadcasting*, London: Paul Elek, 1975.

Herman, Edward S. and McChesney, Robert W., *The Global Media. The New Missionaries of Corporate Capitalism*, London: Cassell, 1997.

Ferjani, Riadh, 'Usages des nouvelles technologies de l'information et de la communication en Tunisie', *Revue tunisienne de communication*, no. 32, 1997, pp. 25–41.

Ferjani, Riadh, 'Internationalisations du champ télévisuel en Tunisie', in Mattelart, Tristan, ed., *La mondialisation des médias contre la censure. Tiers Monde et audiovisuel sans frontières*, Paris-Bruxelles: Ina-De Boeck, 2002, pp. 156–175.

Lamloum, Olfa, *Al-Jazeera, miroir rebelle et ambigu du monde arabe*, Paris: La Découverte, 2004.

Lamloum, Olfa, *La restructuration de l'espace radiophonique arabe. Palestine, Liban, Syrie, Jordanie et Égypte*, Paris: Institut Panos, 2006.

Madani, Lotfi, 'L'antenne parabolique en Algérie, entre dominations et résistances', in Mattelart, Tristan, ed., *La mondialisation des médias contre la censure. Tiers Monde et audiovisuel sans frontières*, Paris-Bruxelles: Ina-De Boeck, 2002, pp. 177–210.

Mattelart, Tristan, *Le cheval de Troie audiovisuel. Le rideau de fer à l'épreuve des radios et télévisions transfrontières*, Grenoble: Presses universitaires de Grenoble, 1995.

Mattelart, Tristan, 'Transboundary flows of Western entertainment across the Iron Curtain', *The Journal of International Communication*, vol. 6, no. 2, 1999, pp. 106–121.

Mattelart, Tristan, ed., *La mondialisation des médias contre la censure. Tiers Monde et audiovisuel sans frontières*, Paris-Bruxelles: Ina-De Boeck, 2002.

Nordenstreng, Kaarle and Schiller, Herbert I., eds, *National Sovereignty and International Communication*, Norwood: Ablex, 1979.

Nordenstreng, Kaarle and Schiller, Herbert I., eds, *Beyond National Sovereignty: International Communication in the 1990s*, Norwood: Ablex, 1993.

Sakr, Naomi, *Satellite Realms. Transnational Television, Globalization and the Middle East*, London: I. B. Tauris, 2001.

Schiller, Herbert I., *Mass Communications and American Empire*, New York: Augustus M. Kelley Publishers, 1969.

Schiller, Herbert I., *Communication and Cultural Domination*, White Plains: International Arts and Sciences Press, 1976.

Scott, James C., *Domination and the Arts of Resistance. Hidden Transcripts*, New Haven: Yale University Press, 1990.

Semelin, Jacques, ed., *Quand les dictatures se fissurent. Résistances civiles à l'Est et au Sud*, Paris: Desclée de Brouwer, 1995.

Semelin, Jacques, *La liberté au bout des ondes. Du coup de Prague à la chute du mur de Berlin*, Paris: Belfond, 1997.

Sreberny-Mohammadi, Annabelle and Mohammadi, Ali, *Small Media, Big Revolution. Communication, Culture and the Iranian Revolution*, Minneapolis: University of Minnesota Press, 1994.

Sreberny-Mohammadi, Annabelle, 'Introduction', in Braman, Sandra and Sreberny-Mohammadi, Annabelle, eds, *Globalization, Communication and Transnational Civil Society*, Cresskill: IAMCR-Hampton Press, 1996, pp. 1–19.

Sreberny-Mohammadi, Annabelle, 'The Media and Democratization in the Middle East: The Strange Case of Television', *Democratization*, vol. 5, nr. 2, 1998, pp. 195–197.

Tomaselli, Keyan and Tomaselli, Ruth, 'Between policy and practice in the SABC, 1970–1981', in Tomaselli, Ruth, Tomaselli, Keyan and Muller, Johan, eds, *Broadcasting in South Africa*, London: James Currey, 1989, pp. 84–138.

Vittin, Théophile, 'Les médias externes comme facteurs de renforcement de l'opposition interne au Bénin (1987–1992)' in Semelin, Jacques, ed., *Quand les dictatures se fissurent. Résistances civiles à l'Est et au Sud*, Paris: Desclée de Brouwer, 1995, pp. 131–150.

GHANIA MOUFFOK

Representation, Images and Censorship in Algeria

How should one refer to the violence passing through Algeria since the nullification of the legislative elections that were won by the Islamic Salvation Front (FIS) in January 1992? As a 'civil war', 'an Islamist rebellion', part of the 'war on terrorism' or 'national tragedy', the current official expression since the election of President Bouteflika in 1999?

How many victims fell in this conflict between 1992 and 2002? Fifty thousand, the number advanced by Ahmed Ouyahia when leading the government, or 200,000, the number suggested by Bouteflika?

This number of 200,000 victims proposed by the larger international Human Rights NGOs has been circulating since at least 1997, already for more than a decade. In the meantime there were other victims, but this number still remains the highest total, as though nobody bothered to count any more. But were the victims of the Algerian drama ever counted?

Every time I write an article about Algeria, or more accurately, every time I write an article, this type of question comes up: what to call things? Which words should one use to account for not only the parts of the Algerian drama about which there is no consensus, but also that which engages my journalistic responsibility, my own ethics?

Depending upon whether I write for publication in Algiers or abroad, I answer that question in a different manner. In both cases my answers betray my opinions; opinions which I made while exercising my profession, without ideological bias – at least, I hope so – unless it is attached to what is called 'human rights', even if war makes a mockery of morals.

For example, during 'the dark decade', for a foreign newspaper I would have written without hesitation 'the civil war', while for an Algerian newspaper I would have written 'violent clashes'. In this plural I would have included state violence (torture, kidnappings, arbitrary justice). Without naming it, I would thus circumvent censorship, because in official terminology there is only one form of violence, that of the Islamists, and even here I hesitate to use this term, since they would have recommended I write 'terrorists' or 'criminals' instead.

This battle over words forcibly divided Algerian and foreign journalists after the higher cadres in the army hierarchy aborted the electoral process. They removed President Chadli Benjedid from office on 11 January 1992 in the name of the defence of 'institutions' and of 'democratic process' against the Islamist project of the Islamic Salvation Front (FIS). The FIS had just been elected by 3.5 million Algerians in the first round of the first legislative elections since independence in 1962. This being the case, must one talk of a coup d'état or of a constitutional void that the army filled only to fulfil its duty?

During these days of anxiety the French-language daily *El Watan* wrote, 'If, during these last days, anyone had thought of preparing a military coup d'état after the results of the first round (...) nothing of the sort happened. Chadli followed constitutional procedures'. Ali Yahia Abdennour, then-president of the Algerian league for the defence of human rights (LADDH), revolted and declared, 'The brutal and unexplained interruption of the democratic process, through the nullification of the election, needs to be analysed as a military coup d'état which has led to the suspension of the constitution.'

This war of words settled in durably like the impending drama that would turn the Algerian media landscape upside down. At stake was the control of news and information and therefore the representation of the protagonists. This should establish the line between 'good' and 'evil' at the cost of true representation of the complexity of the events. In Algeria and elsewhere, writing about an armed conflict is always walking into a minefield, first because of the proximity of death, your own and that of others, and second because each one of the protagonist wishes to impose his version of the facts. The exercise is even more difficult when all of the armed participants hide their identity, which is the case in Algeria. For a journalist it is in fact as difficult to get interviews from leaders of the Islamist rebellion as from a general directing the fight 'against terrorism'. And because the daily news is terrified by emptiness and needs images, without which a conflict does not become 'incarnate', the figure of the 'Algerian journalist' became a symbol, one of the faces of the commencing war. This also made him into a major actor in the media conflict, where his role became a topic of debate. Very soon journalists became targets of violence and around sixty journalist and associates paid with their lives. On 26 May 1993 Tahar Djaout, journalist and writer, was severely injured by a bullet in the parking lot of his building. He died one week later. The funeral in the village of his birth transformed into a large demonstration 'against terrorism'. For part of the public opposed to the Islamist project, journalists became heroes; for the sympathisers of the Islamist cause they were nothing but 'traitors' supporting 'the junta in power', a term employed in the literature of the Islamists at war. In *Etre journaliste en Algerie*, I wrote in 1996:

> The Algerian society demands too much from its journalists, especially from the written press. As if writing gave them some kind of magic power over the events; everybody hoping to see their truths, their expectations and their hates. Newspapers became the only voices, serving as parliament, forum. Journalists are at the same time sociologists of a drama, witnesses of the moment, and activists of their cause ...[1]

For the Algerian army, which in those days represented the state, things were simpler; the role of the journalist was to be at its service. 'Because that is the rhetoric that pleases the state', writes Robert Fisk in *The Great War for Civilization*, 'they want to make us see the war as a tragedy in black and white: Good vs. Evil, "us" against "them", victory vs. defeat. But war is not essentially a question of victory or defeat; war is first of all about dying and inflicting death. It is the absolute defeat of humanity.'

It is the role of the state; the whole question is about how resistant the media are faced with this 'absolute defeat of the humanity'.

In Algeria, in spite of the sacrifice of too many journalists assassinated because of their profession and not because they were covering the war, resistance was rather weak. Very soon journalists became the spokesmen of a biased and partial view of the conflict. In order to understand, we must turn to the history of the media in Algeria.

A brief history of the press

Until 1990, Algerian journalists were civil servants in a country where the state owned all the media, television, radio and written press, a state that furthermore reigned over the whole production and distribution process of the news.

The first bill of information, which independent Algeria belatedly adopted in December 1981, states that 'the news sector is a sector of national sovereignty'.[2] At the time the country was ruled by the only existing party, the National Liberation Front (FLN), which based its power on the 'historic legitimacy' that it had earned as the party that directed the Algerian anti-colonial revolution and led the country to independence in 1962.

The Algerian private press as it presents itself today was born from the violence after the riots of 1988. The riots were spectacular as thousands of young people attacked the buildings symbolising the state's power. For the first time in the recent history of independent Algeria, a state of emergency was declared and the army shot at demonstrators. The official victims number 160, while other sources count at least 500. It was a real shock for Algerian society and announced the divorce of those who are governed and those who govern.

Five month later, on 23 February 1989, a new constitution was adopted in a hasty initiative by the Algerian regime in an attempt to maintain command of the country. It established a multi-party system by recognising (article 40) 'the right to create associations of political character'. This formulation in itself shows the ambiguity and the frailty of the efforts for political opening. Fourteen month later, under the government of Mouloud Hamrouche a new law (law nr. 90 of 3 April 1990) abolished the state monopoly on news and dissolved the Ministry of Information and Culture. Until then the ministry had been the grand organiser of the sector that was actually controlled by the political police (the MS, military security) because news was considered to be an issue of national sovereignty on equal terms with defence. The journalist is no longer an activist but is charged with transmitting the right words to the citizens. From then on the law stipulates that a 'Professional journalist is every person that dedicates himself to research, collection, selection, exploitation and presentation of news and information and makes of this activity his regular profession and his principal source of income.'

If technically this law allows the investment of private capital in any type of media, in fact only the written press benefited from these new arrangements. To this day there have been no private radio and television stations, and audiovisual media remained in governmental hands. The whole experience of the democratisation of information was carried by the private written press, which resulted in it having a role of heightened importance in situations of crisis.

In 1990, however, nobody was prepared for the drama that would follow the nullification of the legislative elections a year later, and a private press emerged as competitor of the state press, structured around capital from different origins. One type of press belonged to political parties, one was owned by private investors and another was owned by collectives of journalists that were originally with the old governmental papers. It is in this last formula that one finds all the uniqueness of the Algerian private press. The so-called 'Hamrouche decree' of March 1990 allowed journalists from the governmental press to form a collective and to benefit from two years of salary each, which was considered equal to the seed capital for new ventures of the written press. It was a real revolution for civil servants as journalists had the chance to become the owners of their own newspapers. Thus the big Algerian dailies were born, such as *Le soir d'Algérie* launched on 5 September 1990, followed by *El Watan*, *Echourouk*, etc. Between 1989 and 1991, the newspaper industry as a whole doubled the number of copies printed, from 800,000 to 1,700,000. There were more than 148 publications, of which fourteen dailies competed in the Algerian market, with mixed success as many did not run past three months.

It was the golden age of the Algerian private press, a real adventure, a new apprenticeship in pluralism accompanying civil society movements that organised into parties and associations in preparation for the first pluralist elections in the country. But soon this press would run up against its dependence. Barely on the market, the private press was squeezed in a matrix of state monopolies on three levels: printing, distribution and publicity. The private editors had no other choice but to print their papers in one of the four state printing presses, the only ones available at the time, which also held a monopoly on the importation of paper. The same constraints applied with distribution, which was dominated by the Entreprise Nationale des Messageries de Presse (l'ENAMEP), and with publicity, which was vital, given the weakness of the seed capital of these new publications. The principal advertiser was again a state enterprise, Agence Nationale d'Edition et de Publicité (ANEP). The market for publicity obeyed no rules of transparency and advertisers chose their support more due to the editorial line than the audience and the volume of the papers. On 20 February 1995, *La Tribune* complained, 'The market for advertisement of the public sector spread over the different titles does not hide the fact that the state tries to bring to heel the publications that do not stand in line.'

All these constraints made it difficult, and that is an understatement, for the new editors to remain independent of the state. The fragility of this press intensified with the halting of the electoral process in January 1991. On 9 February 1992, the state of emergency was declared and in the name of 'the restoration of the public order' all liberties were suspended.

The nullification of the first round of the legislative elections won by the FIS meant in reality the suspension of the democratic process. Algeria no longer had a parliament, political parties were under heavy surveillance and divided, and television and radio became again the loudspeakers for the propaganda of the Algerian authorities that were peacefully rejected at the ballot boxes. And even if the FLN remained the second largest party in the country, they had to face a veritable armed rebellion led by the Islamists. Due to the circumstances, the written press remained the last witness to the aborted democratic process and the only space of pluralist expression. The FIS remained absent here and the party was dissolved on 4 March that year.

The brutal interruption of the elections was experienced by the majority of the population as a profound humiliation. Television and radio remained almost exclusively in the service of the regime and its strategies. They did not feel it incumbent to testify of this grave injury. The private press, on the other contrary, in existence since only two years before, benefited from the generous sympathies of its readers who expected it to tell *al-haq*, an Arabic term meaning faith, right, law, truth and justice.

However, the majority of the private written press was subject to too high a pressure and dependency and did not fulfil these expectations. Solicited by all the political players, those acting in the open as well as those manipulating networks of influence, the press was divided into different camps between the opposition and its solidarity with the population hoping for change, and the suffocating ties that linked them with the state upon which their existence ultimately depended. Algerian journalists were divided like the rest of the country in the face of this new question, 'what is democracy when, for the first time in history, an Islamist party gains legislative power through the ballot boxes?' The question was even more pertinent because the victory of the FIS was overwhelming, with 188 of 231 possible seats in the first round.

It is undoubtedly the chief editor of *El Moudjahid*, the leading newspaper and symbol for the press that works in service of the government, who made the most sincere confession of this rupture. In an editorial published on the front page, he writes:

> In this type of situation it is not simple, and even less easy, to position oneself with regard to the events. Such a choice can only be taken after consulting with the patriot that one is and the democrat that one wants to be. The antagonism follows from the first according priority to the general interest of the nation, while the second accords supremacy to the constitutional laws. In this inner conflict, whatever the moral repercussions on the being, the man and the citizen can only opt for the general interest.

Rereading the news of the terrible days that dragged Algeria into a bloody adventure, one can say that every Algerian journalist is possessed by this exhausting schizophrenia, torn between the 'patriot' and the 'democrat one wants to be.'

Soon, the Algerian regime put an end to these debates by sweeping up all publications that made a choice other than that of 'being patriotic'. At first it banned the newspapers close to the Islamists and later those close to the democratic currents that opposed the nullification of the elections. For the record, one needs to remember that even the FLN (the former ruling party) of Abdelhamid Mehri declared itself to be '... determined to respect the will of the people and to support the democratic process, notably by participating in the second round of the legislative elections' and to warn that 'all other approaches in whatever shape, would be a terrible choice and a real threat for the normal evolution of the society and the stability of the country.'

For its part, the oldest opposition party, the Socialist Forces Front (FFS) lead by Aït Ahamd called for 'respect of legality to avoid a civil war'.

It would thus be wrong to believe that in this crucial period the entirety of the democratic forces in Algeria supported the nullification of the election decided upon by the Algerian army, which pushed the President Chadli Benjedid to step down. It is in this context that the remainder of the private press became the 'last asset of the democratic process'. This paradoxical situation eventually led to the overemphasis of the role of the private press and weakened the profession of the journalists notably when it became increasingly dependent on the state because, in addition to material support, it also required state protection against attempted assassinations.

From this point in time the representation of the conflict by what remained of the private press was dictated by the Algerian army, which had appropriated the means to control 'information related to security matters'. This expression summarises what became of the coverage of the conflict: a simple operation of maintaining order and reinstalling security, threatened by 'evil forces' and fought for by 'the good ones'.

In case anyone did not understand, the Ministry of the Interior and the Ministry of Information and Culture instituted by ministerial decree 'a communication cell to take care of the relations with the media in matters of information, elaboration, and diffusion of official communiqués related to the security situation'. This decree was not 'published' in print but extracts of the provisions it contained were orally communicated to all persons physically and/or morally concerned. Afterwards the private press and the authorities were bound by this secret. This confidential decree was a veritable guide to 'good coverage' of the conflict. Rereading it today, one is above all stunned by the concern accorded to the representation of the Other, the adversary.

The image of the Other

According to the decree, the Ministry of the Interior expected from the media a veritable counter-propaganda mission, 'in order to warn, counter and defeat dissident rumours and propaganda'. It was equally considered necessary to 'reduce the psychological effects that the commanders of terrorism aimed at'. Like soldiers, journalists were expected 'not to panic, to show self-control and determination not to let political violence win'. In the chapter on recommendations one can read that the 'appropriate terminology will be provided to the media by the communications cell'. In the same chapter a robot-portrait of the adversary was proposed under the title 'fight against the ideology and the propaganda of the adversary'. The press was asked to show 'no pictures', not 'to gratify him with unwarranted qualifiers or titles.' But above all it was asked 'to highlight the deception and the dishonesty of those who, in the name of religion and of the purification of society, give themselves to criminal practices, drugs for the executers of terrorist crimes, ex-convicts and bandits used as paid assassins, forced conscription of the youth, cowardly political practices that send the credulous youth to their death.'

For those who nevertheless were tempted to head towards the Maquis, the press was asked to frighten them with the information that 'no crime goes unpunished', highlighting 'the efficiency of the security forces who, even if they do not manage to prevent every crime, will always find the culprits' and emphasising 'the severity of the sentences given by the special court'. By the same logic, the press was asked to provoke reactions of 'rejection of terrorism' by publicising 'the inhuman nature of barbaric practices of terrorism'. Nothing was left to chance or to the initiative of the journalists, who were called to 'emphasise the collusion with foreign powers', to develop a 'collective self-protection reflex', and even to 'put pressure on Algerian clerics, who out of fear remain mute in the face of the terrorist crimes, so that they would take courage to express themselves'.

Unfortunately all these recommendations were implemented to the letter and the slightest overstepping of the secret rules was severely punished, with anything from simply banning the newspaper to a summons to appear before the journalists' tribunal. What was at stake was the population, and the prime objective of the recommendations was not so much the legitimisation of governmental actions, but the de-legitimisation of the adversary, his dehumanisation. Violence was no longer an expression of political failure but a pathology that hence needed to be 'eradicated' by all means, including the most reprehensible. 'You can't make an omelette without breaking a few eggs' was the explanation for what happened to all of those who tried to question the methods of the secret services; summary executions, kidnappings, torture, trials in courts of exception that became the rule etc ... All those who worried about these severe attacks on human rights were mocked with the expression 'the human rightists' ('*les droits de l'hommistes*'). 'If the Algerian press played a central role during this period, it is because it did not content itself with counting the blows and complying with the catastrophe, as the human rightists and other adepts of arithmetical democracy would have liked' – as Abderrahmane Mahmoudi later explained in his book, *Presse algérienne : les nouveaux boucs émissaires* [Algerian Press: the new scapegoat].

Men? Which men?

The figure of the 'terrorist', as the Algerian authorities represented him, is by no means exceptional. 'Terrorism has always been depoliticised by its enemies, criminalised and interpreted as a moral episode, incarnating the impolitic of evil.'[3] 'The terrorist' justifies 'state terrorism', in Algeria as elsewhere.

The representation of the conflict, then, becomes one of the weapons of war, its propaganda. However, the terminology used reflects on the future, on the impossibility of imagining non-violent political solutions. It reflects the state of emergency as the only useful permanent policy in confrontation with an enemy that cannot be defined, an enemy 'without a picture', and hence without a history. Showing a picture of him would show that he resembles a man, a neighbour for instance, and thus would force us to ask questions: Why does my neighbour commit such horrors? Why does he massacre children? Indeed it is true that – according to the language used in the Algerian tribunals – he massacres, rapes and 'sows the fear of desolation'. However, the permanent state of emergency also means restricting all liberties.

From terrorist to kamikaze

Today, violence is no longer as intensive. Algeria has engaged in a policy of 'national reconciliation'; the terminology has changed. After the compromises found between yesterday's adversaries – members of the Islamic rebellion as well as the higher ranks of the army – the all-encompassing phrase 'national tragedy' now summarises all the horrors. A national tragedy that remains full of holes: what happened? The law on national reconciliation, voted for in a popular referendum, forbids this kind of question. It remains forbidden to try to understand, to think and to question the recent past and thus to understand what is going on at the moment.

Until reality, this stubborn thing, catches up with us when human bombs explode in the heart of the capital, Algiers. On 11 April 2007 a suicide attack targeted the governmental palace; the same day another suicide bomber targeted a police station in Bab Ezzouar, the eastern suburb. Together both attacks killed seventeen, and injured 107.

On 11 December 2007, the same scenario repeated itself, this time against the headquarters of the United Nations and the Constitutional Council in Algiers; the toll, thirty-seven dead and 177 injured. 'The terrorist' has turned into a 'kamikaze'.

When his actions are spectacular, and the number of his dead again bespatters the television screens of the planet, he will feature on the front page of Algerian newspapers and remain 'a criminal', 'a junkie', 'manipulated'. The new terrorist is born, but nobody saw him coming.

After an attack against the coastguards on 8 September 2007 (thirty dead, forty injured) in Dellys, a charming little port town on the eastern coast, one learns that the kamikaze was a fifteen-year-old high-school student. His family is described as an average Algerian family who saw in their son a quiet and tranquil boy, loved by all and without any particular problems. He was born in 1992, the year that saw the nullification of the elections won by the FIS, a party he never knew, just as he never knew repression or torture. So where did he find this desire to kill himself and others at the same time?

He might foreshadow a new generation of 'terrorists', a globalised generation whose consciousness is no longer based on a national but on a global history; a generation of children of September 11, rather than October 8, nourished by the war on Iraq, occupied Palestine, exploding Afghanistan and forgotten Chechnya.

This is a generation with internet and satellite dishes that, from the roofs all over Algeria, sends us the chaos of the world. A world in which the United States, the archetypically democratic power, with its big newspapers, CNN, etc, uses the same methods as the authoritarian states, methods believed to be reserved for our underdeveloped countries, our dictators. Saddam Hussein and expeditive justice, Guantanamo, Abu Ghraib, torture, summary executions, no-law zones, military jurisdiction ... a world like Algeria. This generation has the same disjointed vision of reality, the same images of unheard-of violence, the same definition of Good and Evil, the same biased and partial ideas, the same incomprehension about what is going on ...

In Algeria as elsewhere, propaganda and disinformation have disastrous effects on people's consciousness. They prohibit the construction of non-violent alternatives. They are in fact active stakeholders in what they pretend to fight – terrorism.

Notes

1. Ghania Mouffok, *Etre journaliste en Algérie: 1988–1995*, p. 13.
2. Journal officiel no. 6 of 9 February 1982.
3. As Francisco Naishat writes in issue no. 24 of the *Naqd* journal, published in Algiers, and dedicated to terror/ism, state and society.

Bibliography

Fisk, Robert, La grande guerre pour la civilisation, Paris : la Découverte, 2005.
Mahmoudi, Abderrahmane, Presse algérienne : les nouveaux boucs émissaires, Paris : Editions de Poche, 2000.
Mouffok, Ghania, Etre journaliste en Algérie : 1988–1995, Paris : Editions Reporters sans frontières, 1996.

MAHA TAKI

Beyond Utopias and Dystopias: Internet in the Arab world

The Internet and its potential effects on society, politics and the world in general have been widely debated in academic and popular literature in the last ten years. With regards to the Arab world, most vigorously debated is the question of state censorship and the Internet's ability to undermine that censorship. Direct state censorship is indeed a dangerous and important impediment to taking advantage of the Internet in some, though not all, Arab countries. However, we need to look at inequalities in a wider theoretical, historical and cultural context if we are to understand its use and potential impact. There are other important variables, beyond a country's political system, that can determine how technologies are used and appropriated by people, and that can empower or disempower them. This article will look at censorship in a wider context that includes all impediments to gaining access to the Internet that individuals encounter in the Arab world. Inequalities in access to the Internet are all forms of censorship whether they are caused by economic, political or social structures.

Most literature on the Internet fits into alternating dualisms between utopian and dystopian claims. Enthusiasts claim that as the Internet offers decentralised, interactive, non-hierarchal and anonymous communication, everyone should be free to communicate and access information without restriction. Others are sceptical, claiming that the Internet is similar to other media technologies that have done nothing to decrease the knowledge gap. The 'increasing knowledge gap hypothesis' says that as the amount of information in a community or society increases, the gap between the information have and have-nots increases and those with higher socio-economic status are always advantaged in new sources of information.[1]

Most debates that have addressed the Internet have done so in a technologically deterministic manner. This is problematic because it assumes that technology is separated from society and that it is something autonomous that directs and determines social relationships and organisations. Technological deterministic rhetoric also often fails to consider the historical, political and cultural realities against which new technologies should be viewed. We cannot understand how a society or individuals use the Internet and its potential impact on them unless we understand that it is embedded in cultural, symbolic and historical contexts of various kinds.

Researchers concerned with inequality have coined the term 'digital divide' to define the effects of exclusion from new information technologies. The problem with the digital divide is that it is mostly constrained to binary statistics. According to the Internet World Stats Website,[2] in Lebanon, Internet penetration is 20.8 per cent, Jordan 14.8 per cent and Syria 7.7 per cent. Internet penetration figures by country, region or even community measure the simple criterion of access, usually in the convenient locale of one's home and therefore are a reflection of resources ownership such as subscription to an ISP.[3] Deborah Wheeler[4] has continuously pointed out the importance of Internet cafes for Internet browsing in the Arab world. This is mainly because it is far too expensive to have an individual account in one's own home. She believes that connectivity may be higher in the Arab world than conventional figures suggest because of the large numbers of users who use Internet cafes or public access points.[5] Indeed, the average number of users per account can range from 0.6 in Qatar to 38.8 in Sudan, which can reflect the affordability of Internet services, average family size, the number of public access points and so on.[6]

While such statistics could be useful for data gathering and comparison, they do not tell us what happens after access has been achieved.[7] Those with a dial-up connection, for example, will spend less time browsing because it costs more money than does broadband and paying a fixed fee for unlimited access. Factors such as how long it takes to obtain Internet access, who can afford it, how fast and effective it is and whether it is available throughout the country are only some of the variables that need to be considered. There is a distinction between 'formal access', that is physical availability, and 'effective access', that is affordable connectivity and diffusion of skills that people need to benefit from the technology.[8] Questions on whether people know the full extent of the possibilities available for them on the Internet, whether they are aware that they can publish their own content and in their home language, whether they can evaluate the quality of the information they are getting and whether they are taking advantage of the searches on the Internet, are all issues one needs to consider when measuring access and inequalities. Variables such as education and cultural acceptance of technologies are just a few that have also been pertinent in maintaining and increasing inequalities.[9]

Indeed, the social milieu of a person will not change with the introduction of a new technology. Therefore, it is imperative not to lose sight of the wider patterns of social life, history, one's habitus,[10] structural impediments and the contexts in which we are researching. Bourdieu argues that our sensibilities, our sense of beauty, disgust and appropriateness are all formed in a social-historical context of group divisions and conflict.[11] Human beings, for Bourdieu, are cultural beings. We make conscious choices in terms of the way in which we understand our world unconsciously and the meaning it has for us – in other words, we are products of our culture or habitus. It is important to understand that people make conscious choices based on structural unconscious choices that then translate into action. Therefore experiencing life in a certain context involves a habitus which directs our actions.

Internet ethnographers Miller and Slater found in their study on the Internet in Trinidad, that its good integration rate is not only attributed to its healthy economy and good infrastructure but also to the importance of email for communicating with relatives who have gone abroad for work.[12] Fandy states that when radio was introduced to the Arab world, whether private or state-owned, most people listened to it in cafes rather than in the privacy of their own homes,[13] an engagement that was not witnessed in the West. This is because most people could not afford to buy their own radio sets – an economic structure that brought about different consuming habits of a technology.

Internet use in most of the Arab world is restricted to a privileged few. The users constitute a small, well-off and highly educated and young minority. In the majority of the Arab world, Internet use is hindered by high costs and a small percentage of the population owning computers, while in contrast, the average blogger in the United States is reported to be a teenaged girl writing about her everyday life. A study conducted in 2005 in Lebanon, Syria and Jordan revealed that the majority of those who had blogs were young (over 75 per cent between nineteen and thirty), male (64 per cent), well educated, and that the majority had a bachelor and many a master's degree.[14] Furthermore, over half of the respondents had studied in Europe or America.[15] Infrastructure and economic censorship are another problem. Lebanon, one of the few Arab states that is free from the reign of censorship and regulation on the Internet, is burdened by economic and political corruption. It only recently signed a memorandum with a private company to launch ADSL Internet service, rendering it the only Arab country that did not use the ADSL service before 2006. The process of getting it is still very expensive; needless to say, it is only available in central Beirut. A slow and expensive Internet connection can hinder people's searches online and their ability to publish, and the process can be extremely frustrating, resulting in a completely different experience and usage of the Internet from other parts of the world with a fast and reliable internet connection.

This is why, like James Slevin,[16] Miller and Slater,[17] Wheeler[18] and Christine Hine,[19] I argue in favour of a more contextualised, culturally sensitive analysis. Most studies that focus on non-Western usage are framed in terms of access and development. While such studies are valuable, access does not tell us how and why people are using the Internet, and 'development' almost always assumes a benefit. Technologies are not value-neutral but still have both beneficial and disadvantageous consequences.[20] Much of the failure of early promises that communication technologies would assist Third World development can be attributed to the lack of contextualisation.[21] Failing to consider the historical, political and cultural realities and processes against which new technologies should be viewed will only lead to a partial analysis of the Internet and its integration into different societies. Context can help us understand why we see different patterns of usage in the Arab world than in other parts of the world. It also helps us understand where power is concentrated in every society. Assuming that there will be instant shifts in power and authority because the Net is a many-to-many decentralised technology may be ignoring the importance of social structures and historical processes.

The July 2006 war on Lebanon is a good example of how power can shift from the offline world into the online world and depicts how there are some people who are both better equipped to take advantage of ICT and also to use it for the protection of their privileged position; a social and economic process which has much in common with previous epochs.[22] From the very onset of the war, Lebanese people both in Lebanon and abroad were very active in getting messages across, often in real time, via blogs and email, about what was happening. This was reported to be the best example of the revolutionary bottom-up aspect of the Internet. However, in response to the increasing online presence of Lebanese people all over the world, the Israeli foreign ministry retaliated through an online war. The World Union of Jewish Students (WUJS) introduced megaphone software that could be downloaded from the 'Give Israel your united support' website.[23] The Israeli foreign ministry ordered trainee diplomats, who were later named cyber soldiers, to track websites and blogs with this software, so that networks of US and European groups with pro-Israeli activists could place supportive messages. The cyber soldiers would move from one poll to another and one blog to another and write one-off replies to posts that were deemed to defame the Israeli state.[24] Anyone could sign up for the software (and still can today) from the giyus.org website and be alerted to any anti-Israeli rhetoric on the Web so they can post the contrary view. Their motto is 'Today's conflicts are won by public opinion. Now is the time to be active and voice Israel's side to the world.' To attract users from around the world, giyus.org translated polls into English, French, Hebrew and Spanish.[25] The Israeli government's interest during the summer war showed how seriously they were taking this new communication medium and that they were ready to invest in and spend resources to try to intimidate the public with rhetoric and propaganda. It also showed that the Internet does not present a picture of a new virtual world separated from the political, social and economic concerns of the old 'real' world order.[26]

In the Arab world, government censorship is still rife. The methods governments use are often ambiguous and random. They can be very direct in their actions, from blocking certain sites to intimidating, threatening and sometimes imprisoning Internet users and bloggers who deviate from the accepted status quo. There can also be more nuanced forms of censorship, based on surveillance that can lead to self-censorship. Some countries have even started to restrict the use of anonymous identities which many bloggers use to circumvent restrictions. *Al Safir* newspaper[27] and the *Malaysia Sun*[28] reported that the Syrian ministry of communication and technology has taken measures to restrict the use of anonymous comments. In a decree issued on 25 July 2007, the government required all website owners to display 'the name and e-mail address of the writer of any article or comment [appearing on their site] ... clearly and in detail, under threat of warning the owner of the website, then restricting access to the website temporarily and, in the case that the violation is repeated, permanently banning the website.'[29]

While the direct application of such measures may not necessarily affect the average user directly – that is, the Syrian government will not be able to track down every single anonymous comment or website – nevertheless, the perception that the government or society as a whole is engaged in the surveillance and monitoring of Internet activity, whether accurate or not, can lead to an internalisation of these types of structures and thus, self-censorship. In spite of everything, those living in controlled-media countries are more likely to be afraid of the written word and are less likely to blog about politically and socially sensitive issues than those who have 'free media' systems. Many Lebanese bloggers have reported that they self-censor, especially political rhetoric, because they do not want their friends to know their opinion or to offend or upset those they know, especially if they are deviating from what the rest of their sect or community thinks about a certain issue. This is particularly true in today's political environment. Indeed, a survey conducted in 2005 on bloggers in Lebanon, Syria and Jordan revealed that 80 per cent of the ninety-one bloggers who responded stated that they practise self-censorship.[30]

The impact of this direct form of censorship is ambiguous. Those who have the resources and are determined will find a way to circumvent these restrictions – from buying some software to using servers in other countries. A Human Rights Watch report[31] depicted most Syrians, for example, as dialling Lebanese or Jordanian ISPs to get past their government censorship. This is one of many ways that Syrians and others can bypass restrictions. In fact, many initiatives on the Web can be found that give advice to users on how to be anonymous and get past filtering and blocking. The question remains, however, how many Internet users will go to such measures to be able to access information or publish online? Can everyone afford to buy software that allows users to be anonymous and to bypass firewall restrictions? Are people aware that it exists? How expensive is it and how does word spread about it? It appears that those who will make an effort to get round the censorship are just the 'techie' people and those with very high motivations for using the Net, such as political opposition, gay people and other oppressed groups. Will people not bother and would rather just self-censor?

Self-censorship is not necessary a conscious decision resulting from an overt external force. Self-censorship is often a decision that we are unaware of, stemming from these systems of power – all, in a way, related to the making of a specific social world. The perception that a society is watching can lead to this form of internalisation of surveillance and thus to self-regulation and policing. The relationship between systems of social control and people in a disciplinary situation is accurately described by Michel Foucault,[32] who states that the result of surveillance is acceptance of regulations and docility – a normalisation of sorts – stemming from the threat of discipline. Disciplinary power, according to Foucault, works by making those subject to it visible, as in the example of the Panopticon. Subjects in turn internalise this surveillance. As Phillip Agre argues in his paper 'Internet and the political process', the culture has to want the freedom given to it, and in many societies authoritarian habits, beyond a narrow stratum of intellectuals, run deep.[33]

The Internet does indeed create many opportunities in the Arab world and has played a positive role, especially amongst opposition groups and minorities. Its impact however, is neither utopian nor dystopian. Like many social spaces, it is a contested terrain and a battleground of discourses.[34] It is simplistic to presume that there will be instant shifts in power and social structures because of it. There are still inequalities shaped by remaining traces of the context we live in and our past experience. The digital divide is a political outcome rooted in these historical systems of power and privilege and not simply a gap in access to and use of the Internet and computers.[35] Therefore, to understand its impact, we have to understand the political system as one structuring element that influences people to make choices in a particular way, along with the economy, the social environment and habitus.

Notes

1. J. Robinson, P. DiMaggio and E. Hargittai, 'New Social Surveys Perspectives on the Digital Divide, *IT & Society*, no.1, issue 5, Summer 2003, p. 2.
2. [http://www.internetworldstats.com/].
3. P. DiMaggio, E. Hargittai, W. R. Neuman and J. Robinson, 'Social Implications of the Internet', *Annual. Review of Sociology*, no. 27, 2001, p. 311.
4. D. Wheeler, *The Internet in the Middle East: Global Expectation and local imaginations in Kuwait*, New York: State University of New York Press, 2006.
5. D. Wheeler, 'Blessings and Curses: Women and the Internet Revolution in the Arab World', in Sakr, Naomi, *Women and Media in the Middle East: Power Through Self-Expression*, London: I. B. Tauris, 2004, p. 138.
6. B. Warf, and P. Vincent, 'Multiple geographies of the Arab Internet', *Area*, no. 39, issue 1, 2007.
7. DiMaggio, Hargittai, Neuman & Robinson, loc. cit.
8. E. Wilson, 'Closing the Digital Divide: An Initial Review', *Briefing the President: Washington DC: Internet Policy Institute*, May 2000, cited in DiMaggio, Hargittai, Neuman and Robinson, op. cit., p. 46.
9. F. Magnisi, 'Understanding the digital Divide: Stratification and Empowerment in the Kingdom of Tonga', *PTC proceedings*, 2007.
10. P. Bourdieu, *Outline of a Theory of Practice*, Cambridge University Press, 1977.
11. N. Crossley, *Key Concepts in Critical Social Theory*, London: Sage, 2005, p. 108.
12. D. Miller & D. Slater, *The Internet: An Ethnographic Approach*, Oxford: Berg, 2000 cited in DiMaggio, Hargittai, Neuman and Robinson, op. cit., p. 313.
13. M. Fandy, 'Information Technology, Trust and Social Change in the Arab World', *Middle East Journal*, no. 54, issue 3, 2000, pp. 381.
14. M. Taki, Weblogs, *Bloggers and the Blogosphere in Lebanon, Syria and Jordan: An Exploration*, MA Dissertation, London: University of Westminster, 2005.
15. Ibid.
16. J. Slevin, *The Internet and Society*, Oxford: Polity Press, 2000.
17. D. Miller and D. Slater, The *Internet: An Ethnographic Approach*, Oxford: Berg, 2000.
18. D. Wheeler, *The Internet in the Middle East: Global Expectation and local*

imaginations in Kuwait, New York: State University of New York Press, 2006.
19. C. Hine, *Virtual Ethnography*, California, Sage, 2002.
20. B. Loader, *Cyberspace Divide: Equality, Agency and Policy in the Information Society*, 1998, London: Routledge, p. 6.
21. F. Williams, R. Ronald, E. Rogers, *Research Methods and the New Media*, New York: Free Press, 1988, p. 45.
22. Loader, p. 8.
23. [http://www.giyus.org/].
24. Farago, Yonit, 'Israel backed by army of cyber-soldiers', in *Times Online*, 28 July 2006 retrieved 9 September 2007 from [http://www.timesonline.co.uk/tol/news/world/middle_east/article693911.ece].
25. S. Linde, 'Israel's newest PR weapon: The Internet Megaphone' in the *Jerusalem Post Online*, 8 November 2006 retrieved 11 January 2008 from [http://www.jpost.com/servlet/Satellite?cid=1162378505678&pagename=JPost%2FJPArticle%2FPrinter].
26. Loader, p. 6.
27. W. Ibrahim, 'Syrian Communication Ministry wants the heads of online commentators', *Al Safir Newspaper*, 3 August 2007, retrieved from [http://www.assafir.com/Article.aspx?EditionId=706&articleId=273&ChannelId=15664&Author=????%20??????].
28. Malaysia Sun, 'Syria Under the Hammer over Online Censorship', in the *Malaysia Sun*, 12 October 2007, retrieved 07 November 2007 from [http://story.malaysiasun.com/index.php/ct/9/cid/b8de8e630faf3631/id/290301/cs/1].
29. W. Ibrahim.
30. Taki.
31. E. Zarwan / *Human Rights Watch*, 'False Freedom: Online Censorship in the Middle East and North Africa', New York, November 2005.
32. M. Foucault, *Discipline and Punish: The Birth of the Prison*, New York: Random House, 1975.
33. P. Agre, 'Real-time Politics: The Internet and the Political Process', *The Information Society*, no. 18, issue 5, 2002, p. 317.
34. B. Warf and J. Grimes, 'Counterhegemonic Discourses and the Internet', *Geographical Review*, no. 87, issue 2, April 1997, pp. 270.
35. L. Kvasny, 'The role of the Habitus in Shaping Discourses about the Digital Divide', *Journal of Computer-Mediated Communication*, no. 10, issue 2, 2005, p. 2.

Bibliography

Agre, Phil, 'Real-time Politics: The Internet and the Political Process', *The Information Society*, no. 18 (5), 2002, pp. 311–331.

Bourdieu, Pierre, *Outline of a Theory of Practice*, Cambridge University Press, 1977.

Crossley, Nick, *Key Concepts in Critical Social Theory*, London: Sage, 2005.

DiMaggio, Paul; Hargittai, Eszter; Neuman, W. Russell; Robinson, John P, 'Social Implications of the Internet', *Annual. Review of Sociology*, no. 27, 2001, pp. 307–336.

DiMaggio Paul, Hargittai, Eszter, Celeste, Coral and Shafer, Steven, 'Digital Inequality: From Unequal Access to Differentiated Use', in Neckerman, Kathryn, *Social Inequality*, New York: Russell Sage Foundation, 2004, pp. 355–400.

Fandy, Mamoun, 'Information Technology, Trust and Social Change in the Arab World', *Middle East Journal*, no. 54 (3), 2000, pp. 378–394.

Farago, Yonit, 'Israel backed by army of cyber-soldiers', in *Times Online*, 28 July 2006 retrieved 9 September 2007 from [http://www.timesonline.co.uk/tol/news/world/middle_east/article693911.ece].

Foucault, Michel, *Discipline and Punish: The Birth of the Prison*, New York: Random House, 1975.

Hine, Christine, *Virtual Ethnography*, California: Sage, 2002.

Ibrahim, Wassim, 'Syrian Communication Ministry wants the heads of online commentators', *al-Safir Newspaper*, 3 August 2007, retrieved from [http://www.assafir.com/Article.aspx?EditionId=706&articleId=273&ChannelId=15664&Author=????%20???????].

Kvasny, Lynette, 'The role of the Habitus in Shaping Discourses about the Digital Divide', *Journal of Computer-Mediated Communication*, 2005, no. 10 (2).

Linde, Steve, 'Israel's newest PR weapon: The Internet Megaphone' in the *Jerusalem Post Online*, 8 November, 2006 retrieved 11 January 2008 from [http://www.jpost.com/servlet/Satellite?cid=1162378505678&pagename=JPost%2FJPArticle%2FPrinter].

Loader, Brian, *Cyberspace Divide: Equality, Agency and Policy in the Information Society*, 1998, London: Routledge.

Magnisi, Fonongava'inga, 'Understanding the digital Divide: Stratification and Empowerment in the Kingdom of Tonga', *PTC proceedings*, 2007.

Malaysia Sun, 'Syria Under the Hammer over Online Censorship', in the *Malaysia Sun*, 12 October 2007, retrieved 7 November 2007 from [http://story.malaysiasun.com/index.php/ct/9/cid/b8de8e630faf3631/id/290301/cs/1].

Miller, Danny & Don. Slater, *The Internet: An Ethnographic Approach*, Oxford: Berg, 2000.

Robinson, John, DiMaggio, Paul, Hargittai, Eszter, 'New Social Surveys Perspectives on the Digital Divide, *IT & Society*, no. 1 (5), summer 2003, pp. 1–22.

Slevin, James, *The Internet and Society*, Oxford: Polity Press, 2000.

Taki, Maha, *Weblogs, Bloggers and the Blogosphere in Lebanon, Syria and Jordan: An Exploration*, MA Dissertation, London: University of Westminster, 2005.

Warf, Barney and Vincent, Peter, 'Multiple geographies of the Arab Internet', *Area*, no. 39 (1), 2006, pp. 83–96.

Warf, Barney and Grimes, John, 'Counterhegemonic Discourses and the Internet', *Geographical Review*, no. 87 (2), April 1997, pp. 259–274.

Wheeler, Deborah, *The Internet in the Middle East: Global Expectation and local imaginations in Kuwait*, New York: State University of New York Press, 2006.

Wheeler, Deborah, 'Blessings and Curses: Women and the Internet Revolution in the Arab World', in Sakr, Naomi, *Women and Media in the Middle East: Power Through Self-Expression*, London: I. B. Tauris, 2004, p. 138.

Williams, Frederick, Rice, Ronald, Rogers, Everett, *Research Methods and the New Media*, New York: Free Press, 1988.

Wilson, Ernest, 'Closing the Digital Divide: An Initial Review', Briefing the President: Washington DC: Internet Policy Institute, May 2000, cited in DiMaggio, Paul; Hargittai, Eszter; Neuman, W. Russell; Robinson, John P, 'Social Implications of the Internet', *Annual Review of Sociology*, no. 27, 2001, pp. 307–336.

Zarwan, Elijah/ Human Rights Watch, 'False Freedom: Online Censorship in the Middle East and North Africa', New York, November 2005.

HUSAM TAMMAM

A Reading of the Media Performance of the Muslim Brotherhood in the 2005 Parliamentary Elections: A Case Study of the City of Alexandria[1]

The Internet is the communications medium that has witnessed the biggest boom in the past years in Islamic media, as it is also the widest and most extensive space available to the Muslim Brothers under the legal prohibitions that still pursue them and due to which they are refused a presence in newspapers and on TV.

The last parliamentary elections, in 2005, were the most crucial event on Egypt's political field in past years, as they witnessed the most active movement of the Brotherhood (as compared with other political events, including presidential elections, for example) since the Brotherhood took a full and active part in elections.

Alexandria, the city traditionally known for the Brotherhood's influence, was the most active place with a trend for development, and had a large presence of Muslim Brothers. Then came the second round of elections, which was very decisive, in addition to having the most active media coverage; 70 per cent of the widest coverage of this round was focused on and emanated from the Muslim Brothers. During those elections the Brotherhood was able to win eighty-eight seats, or about a fifth of the Egyptian Parliament.

The Muslim Brotherhood's media experience in recent years should be examined within the context of the boom in independent and alternative media in Egypt. The country underwent a major shift after the rise of political movements and the escalation of the debate over reform initiated by the launch, in December 2004 of the Egyptian movement for change, *Kefaya*. The debate intensified after President Hosni Mubarak decided in March 2005 to amend article 76 of the constitution to allow more than one candidate to run for the presidency in elections held in September 2005. The debate climaxed during the elections, which went on into three rounds in November and December 2005.

This boom featured the publication of numerous newspapers and independent website launched within a short period of one another in 2004 and 2005, having in common – despite the different ideologies – an interest in internal political reform and giving priority to internal affairs over regional and international issues. Among the main newspapers published at this period that spearheaded this trend were the *Nahdat Masr* (Egyptian Renaissance)[2], *al-Masri al-Youm* (The Egyptian Today)[3], *al-Dustour* (The Constitution)[4], and *al-Mesryoon* (Egyptians), an e-newspaper which was not authorised to publish in print.[5] All these newspapers, as indicated by their names, reflected a growing interest in internal affairs that almost overshadowed the Arab and international issues which had always been strongly present in the Egyptian media.[6] This appears to be a new phenomenon, particularly for Islamic movements, as shown by the *Egyptians* e-newspaper, published by Islamists close to the political Salafi movement.[7]

Furthermore, the boom encompassed personal blogs on the Internet, featuring a great many launched by Egyptian youths who often belonged to neither political organisations nor parties as such, but were interested in reform and seek to declare their stance on reform and the political scene in general. These blogs played a major role in exploding political issues, providing a detailed coverage of election-rigging activities and assaults on political forces and calling for reform. Also exposed was the torture undergone by citizens in police stations, which prompted the filing of some lawsuits against the perpetrators. These blogs had a major influence on the activation of the reform movement and provided a wide and live coverage of its actions.

In this context, the Muslim Brotherhood undertook active steps to develop websites and support existing ones. This movement gathered speed as the Brotherhood participated in the debate on reform and thousands of its members took to the streets in large protests that covered many cities and led to the arrest of hundreds of its members by the authorities.[8] The website development movement climaxed with the opening of candidacy submissions for parliamentary elections and the Brotherhood's declaration of its participation, announcing the constituencies in which it intended to run and the names of its candidates.

As the initial round of parliamentary elections started, the Brotherhood had already expanded to establish a number of additional media committees which, in turn, launched websites covering the constituencies where the Brotherhood was running. In some constituencies, like those of the district of Alexandria, the action further developed into allotting a website for each Brotherhood candidate.

In general, we can distinguish between two types of website that adopt the Brotherhood's stance. The first is more of an official website, a mouthpiece to the Brotherhood and a dispenser of information about it (e.g. www.ikhwanonline.net, the official website of the Brotherhood in Egypt and www.amlalomah.net, the Brotherhood's website in the district of Alexandria). Such websites are the most widespread among Brotherhood websites and always expressly defend its positions, politics and projects. The media material featured on these websites is often directional, defensive and mobilising. The second type of Brotherhood website is argumentative, meditative and sometimes critical in nature (e.g. Brotherhood forums). Such websites are scarce and tend to be individual and private websites developed by some Brotherhood youths. They do not represent the Brotherhood, even though they are committed to its stance and defend the projects of the Muslim Brotherhood.[9]

The Brotherhood's leadership set a strategy aimed at activating media performance via the Internet, thus breaking the media blackout it was subject to. It had predicted the persistence and escalation of the media blackout during the elections, and had had to confront the media campaigns that accompany such events.[10]

Due to the elitist approach to Internet use in a country such as Egypt, the Brotherhood's strategy classified the audience to be targeted by its media activities into the following segments: the Brotherhood's power base and an audience of supporters, devotees and sympathisers, with whom the Brotherhood could communicate safely via this media network; local, Arab and international media, which the Brotherhood could effectively feed detailed and authorised information via the network; and the observational association responsible for supervising the electoral process. The Brotherhood was keen to establish communication channels with this institution to convey testimonials on the elections.

Alexandria[11] was one of the major districts benefiting from a strong media presence from the Brotherhood.[12] It was among the districts that established the highest number of media committees, which included twelve sub-committees covering the entire city, headed by the central media committee. Alexandria was also the most proactive in launching websites and the only district in Egypt that launched a live-coverage radio station via the Internet, Sama, at www.samaradio.com.

The Muslim Brotherhood ran for twelve of the twenty-two parliamentary seats in the district of Alexandria. These were in the constituencies of al-Mountazah (one candidate), al-Raml (two candidates), Bab Sharq (one candidate), Gherbal (one candidate), Qarmouz (two candidates, one of whom withdrew his candidacy), al-Jamarek (two candidates), Mina al-Basal (two candidates) and "al-'Amirya (one candidate).

The Brotherhood officially nominated twelve candidates. But it also chose to unofficially nominate other precautionary candidates in the same constituencies – anticipating any arrests, pressure or emergencies that might hinder the original candidates and their continued participation in the race. This number was considered to be relatively high, covering almost all of the areas in the district.[13]

The Brotherhood established twelve media sub-committees, one for each candidate, headed by the central media committee that was directly linked to the administrative office, the highest administrative authority in the district. These committees were first established on the basis of voluntary contribution, obtaining assistance from members in the Brotherhood alone and limiting outside assistance to a minimum and to areas that needed to be addressed by professionals, thus ensuring the least possible contact.

Hence the Brotherhood made sure most of the time that media sub-committees be made up of members with an interest in working in the media or with related previous experience, specialisation or hobbies. It was also keen to pick Brotherhood sub-committee members from the regions covered by their sub-committees, in order to facilitate their jobs and ensure time-efficient communication between members and the Brotherhood leadership in those regions.

The Brotherhood organised intensive training, administered to the members of these sub-committees by specialised Brotherhood members. Some training courses included sessions in media work, administered in specialised media centres. The Brotherhood also made sure that sub-committees were provided with the appropriate professional equipment, such as video cameras, tape recorders, mobile phones and laptops. The central media committee was given two DV 6m cameras, which can capture footage that may then be broadcast on satellite television channels without processing, in order to facilitate their use of exclusive material provided by the committees.

The organisation also sought to train its members in precision and discretion, securing it against communication network tapping. To this effect, it provided the members with a set of mobile phone lines that had never been used before in Brotherhood activities. A closed and separate network was established between them, insuring them further against security pursuit.

The Brotherhood's central media committee in Alexandria gave the sub-committees the prerogative to make decisions and decentralise operations. This was undertaken to immediately provide a free and decentralised flow of all information pertaining to the coverage of elections, related events and activities. This procedure was further developed into a round-the-clock operation on voting days.

The media committee's work was divided into four major specialities covering all the stages of the media process. First, the public relations committee, which ensured communication with the press, news agencies, satellite television channels, observers and interested rights organisations, and public figures interested in the elections. It also issued and disseminated statements and received feedback.

Second was the training committee, which formed media cadres and ensured the skills-building and development of media committee members through training, development and special programmes. Next was the production committee, producing advertising, printed, audio and video material.

The Internet committee was an independent group within the media committees, focusing solely on the Internet and the dependence of the Brotherhood on this medium, which is the least expensive and most widespread of the available media. The committee developed six websites covering the six constituencies where Brotherhood candidates were running. It also launched the first Brotherhood radio station in Egypt, Sama, broadcast via the Internet at www.samaradio.com.

The media committees managed their activities and varied them according to the nature of the targeted audience: the power base and constituencies of supporters, devotees and sympathisers, media, observers and rights groups.

The media committees started classifying websites according to the constituencies they covered. Each website covered its own constituency, making sure to include all its areas and not to overlap between different constituencies. Each website was also appointed election figures and activities specific to the constituency it represented, in coordination with Brotherhood officials responsible for the elections in that constituency. Among the actions undertaken by each media committee to control its performance and go beyond mere random media coverage to a targeted one (coverage featuring direct and live broadcasts to the targeted audience) was organising a closed mobile phone network within the limits of said area. For this purpose, mobile phone numbers were gathered within the constituencies of Brotherhood members (relatives, friends, neighbours, work colleagues, etc.), thus creating a database for the members of the constituency and ensuring that propaganda for the candidates was sent precisely and regularly.

At the same time, the central media committee was keen on striking a balance between general media efforts and targeted ones. It therefore undertook actions that sometimes involved specific segments of the community and were not based on geographical considerations, such as issuing the Parliamentary Teachers' Bulletin that essentially and exclusively targeted an audience of schoolteachers. The committees were also linked to the other operational bodies within the Brotherhood, which worked on advertising for its candidates, such as Brotherhood-affiliated mosques and charities, and other related circles.

Media committees prepared for actions targeting media, observers and rights groups by creating a database of these organisations and entities, including all the means to contact the committees and their staff, particularly those belonging to the civil society. Media committees also worked on putting together periodic reports on the progression of the elections. These reports increased in frequency on polling day and were issued by the hour, as the committees sent them instantly and periodically to all organisations and concerned parties. The committees also provided live and documented material on many events and facts that were at that stage unattainable by media and observers. In addition, the presence of the committees in almost all of the constituencies allowed them to document and to periodically notify observers of the violations and infractions attributed to some candidates, security bodies or other authorities.

The Brotherhood media committee in Alexandria ensured the professional management of the Sama radio station. It therefore utilised all possible forms in running the station.

1. News reports: they covered the marches, electoral procedures, and provided updates on the race.
2. Live and recorded interviews: these were often conducted with Brotherhood candidates and sometimes with opponents of the ruling National Party and other Brotherhood rivals. The interviews were sometimes conducted with voters who sympathised with the Brotherhood. One of the funniest interviews was with a woman presented by the station as having had a dream in which it was revealed to her that she should support a Brotherhood candidate. Another interview was conducted with the female owner of Hayat's Hairdressing Salon, who by all logic would not be considered among the traditionally committed and conservative Brotherhood supporters, as she explained the way – through her son – in which she had discovered that the Brotherhood is a group of nice people and that she would support them.
3. Drama: this included audio sketches focusing on exposing what were believed to be security infractions and election rigging, or sketches criticising the negative attitude of the electorate towards the elections, albeit in a satirical way.
4. Anthems and songs: the station aired a number of enthusiastic anthems

and songs, some recently produced in Egypt, including the song *Qul Al-Sha'b Qalha Qawyya Nhib Al-Islam* (All the People Arduously Said: We Love Islam) and the anthem *Bihlam Ana Al-Masry* (I Dream, Me the Egyptian), as well as others produced abroad, including the anthem from the Brotherhood in Jordan: *Qal Al-'Akhyar Qalimatahum* (The Good Have Said Their Word).

5. Poetry: the radio aired mocking poems by the Nasserite and Islamic-oriented poet Ameen Al-Deeb, whose poems carry Brotherhood messages. The Muslim Brotherhood produced a cassette tape of Al-Deeb's poems, entitled *Boss W Shuf* (Look and See), conveying biting criticism of the regime, which it distributed widely.

6. Promotions: the Brotherhood enlisted professional actors to star in advertisements depicting a positive picture of the Muslim Brotherhood, complementing its ideas and members and calling for support in the elections. This was considered to be a new and unprecedented means of Brotherhood promotion. The radio aired a number of these advertisements time and time again, while the central media committee in Alexandria broadcast them as candidate propaganda on many satellite television channels such as al-Nass (The People) and SmartWay. They also served as features on the big screens set up in public places and squares during the candidates' electoral campaigns.

All indicators point to the fact that the Brotherhood media committee in Alexandria succeeded in giving a distinguished performance during electoral campaigns. This can be gleaned from the following facts.

1. The Brotherhood was able to organise many successful press conferences, some of which were attended by Arab satellite television channels (such as the Qatari al-Jazeera). The media committee provided live broadcasts of the conferences, the holiday prayers and the large *iftar* held by the Brotherhood for community figures and political powers in Alexandria.

2. The committee also succeeded in scoring media scoops, such as airing a video report on Sama capturing a consultative council (*Majlis al-Shura*) member of the ruling Democratic Party allegedly assaulting

posters and advertisements showing candidates from the opposition with a hammer.

3. The committee developed its technical and media capacities, ensuring thereby the continued airing of Sama radio throughout the day, including eight hours of live airtime during voting days, which then decreased to four hours.

4. An internal Brotherhood report records that its official website, www.amalalomah.com, experienced a record number of hits, reaching 250,000 in November, during which the elections took place, and 35,000 hits on polling day. As for the Sama radio station, it registered 2,100 hits at any given moment, reaching an audience of 14,000 on the day of the vote.

5. Whilst covering the second round of the elections, most Egyptian and Arab newspapers based their stories on reports disseminated by the Brotherhood media committees in Alexandria. A report by the central media committee in the district – the accuracy of which cannot be confirmed – estimated that 70 per cent of the coverage of the *al-Masri al-Youm* daily was essentially based on reports disseminated to reporters and broadcast on Brotherhood-affiliated websites by the media committees.

6. These committees were also able to present distinguished and professional media material which was rebroadcast by other professional media. Dream, a private television network, rebroadcast on its ten o'clock evening show three reports prepared by the Brotherhood media committee in Alexandria. Among such broadcasts was a video report on the protests of the Sidi Beshr church where the committee cameras captured live footage of the protest and the following clashes, as well as a statement from the Brotherhood's MP in the district, Mustafa Muhammad. Another video report featured the protest held by the faculty of Alexandria University, denouncing the arrest by the security forces of university professors who were members of the Brotherhood (Hassan al-Prince and Ali Barakat). The Associated Press news agency also bought the video report on the church incident for $500.[14]

7. Observers, as well as the Arab and international media, were impressed

by the performance of the Brotherhood in Alexandria. Many among them set up venues to analyse this performance and deemed it a distinguished interaction with the media, surpassing the capacities of official, regime-affiliated and National Democratic Party media. This was achieved in the context of the difference in means, the restrictions, tightened measures and the legal ban imposed on the Brotherhood.[15]

This researcher observes that the performance of the Brotherhood media committees became more professional but remained, nonetheless, within the bounds of directional and mobilising propaganda. This type of approach accelerated, while media venues promoting discussion, debate, evaluation and re-evaluation were restricted. Furthermore, the latter type of media is scarce given the rarity of venues sanctioning it and the limited space given to communication within Brotherhood websites and venues.

The official website of the Brotherhood in Alexandria (www.amalalomah.com) lacked, until recently, a space for discussion. The communication venue (the amalalomah forum) was launched after the elections ended. This venue tended at first to be flexible, open and tolerant of change. The forum overseer set three topics for discussion and would not overly interfere with the ongoing debate. However, interference increased gradually, to control the discussion and keep it in line with the general orientation of the forum, which tends not to criticise or evaluate members.

The same applies to Sama radio. Brotherhood officials opened the airwaves to participation for the two months of the elections. Nevertheless, this came to a quick and absolute end, for the alleged reason of profane and unorthodox participation, and from an administrative point of view due to the absence of an available supervisor to control participation.

The Brotherhood's expanded array of websites and the increased demand on Internet services led to important results that went beyond giving it a foothold on the scene, to incite major internal changes within the organisation. These changes served to link the Brotherhood to new media and introduce it to the issue of education and information dissemination.

The demand of new media increased within the Brotherhood, whose staff dealt more often with the Internet, particularly thanks to the optional policy to encourage members. The demand persisted and increased even after the elections. This affected the very structure of the Brotherhood and gave way to a new transformation that, if carried on, could influence the internal formation, and education within the organisation. This influence may be manifested in obliterating the traditional method of information dissemination and education, which depends on a top-down approach and direct contact, transforming it into a new media-dependent approach.[16]

This expansion towards the Internet and the increase in authorised Brotherhood websites changed the patterns of relationships within the organisation. There was a shift from issuing orders directly to sending them through modern media. Organisational orders were thereby transformed into news. News that had traditionally taken a long time to be delivered could now be posted instantly (it takes thirty seconds to launch on Brotherhood websites), and could be sent via e-mail or SMS. This limited the dominance of the education department by restricting its exclusive responsibility to disseminate news. The Internet made organisational information available and reduced the monopoly of department officials. It also served to stave off filtering, made information perhaps more unreliable and maintained relevant anonymity.

After the election the Brotherhood proceeded to expand information on formation and education, and sometimes, even that relating to the organisation itself. This was done via its websites, in order to facilitate retrieval of such information and to save time. The vocational dissemination department started making the Friday sermon available on www.amalalomah.net where the speaker could retrieve it every Tuesday. This step was further developed into an even easier and more universal approach. It changed from a password-only service into an open-access one. Now, speakers could retrieve original texts and edit them as per the requirements of the mosque in which the sermon would be given.

The Brotherhood also set an e-training policy and started giving formation sessions (such as meeting management, self-management, etc.) via the Internet, which reduced lecturer-listener sessions. The Brotherhood also adopted a policy of sending organisational orders by SMS on mobile phones, while making sure not to disseminate them randomly and to use only known and reliable numbers.

This contributed to taking the Brotherhood into the public arena after reducing specialised access. Some Brotherhood websites offered statements from highly placed administrators in the organisation (district officials, administrative office members, area officials) via phones and e-mail.[17]

Notes

1. This paper was discussed at the 'International Conference on Public Opinion and the Media between Europe and the Middle East', organized by L'Institut Français du Proche-Orient, the Orient-Institut Beirut and the Friedrich-Ebert-Stiftung, Beirut-Lebanon, 15–17 November 2006.
2. Liberal daily newspaper launched in 2004, published by Good News 4 You production company owned by media mogul Emadeddine Adeeb.
3. Liberal daily newspaper launched in 2004, published by a private press publishing company of the same name (Al-Masri al-Youm Foundation for Press Printing and Publishing) owned by liberal businessmen, among whom Tawfic Kamel Diab.
4. Daily newspaper launched in 2005 by a private company (Al-Dustour Press and Publishing Company). Chairman of the Board: businessman Isam Ismael Fahmi. Editor-in-Chief: Ibrahim Issa.
5. Launched in 2005 by a group of journalists and writers affiliated to the politically active Salafi movement in Egypt, among whom were the brothers Mahmoud Sultan, website editor, and Jamal Sultan, Islamic writer and editor-in-chief of the *Al-Manar al-Jadid* Islamic intellectual periodical. Recurrent attempts were made to publish it in print but the competent authorities refused to grant it authorisation.
6. The founders and most of the writers of the newspaper belong to the Salafi movement. The newspaper is nonetheless open to the publication of material from all currents. Its editor-in-chief held the same position at *Al-Manar al-Jadid* Islamic intellectual periodical, and is among the key figures of the Salafi reform movement. He also spearheaded the initiative within Salafi circles to form the Al-Islah (Reform) political party, but the regime refused to grant him authorisation.
7. Among the newspapers in this area is the *Al-Ghad* (Tomorrow) newspaper, launched by Al-Ghad Party that is headed by Ayman Nour. Nour was the main opponent of President Hosni Mubarak in the presidential race. We deemed it inappropriate to list the said newspaper above since it is affiliated to a political party.
8. Brotherhood members started taking to the streets by the end of March. This movement escalated in many Egyptian districts and was dubbed the 'spring of independence'. However, the authorities launched an arrests campaign that touched hundreds of protesters. This lead to a deceleration of Brotherhood protests until they finally ceased shortly before the presidential elections in September of the same year.

9. For further information on Brotherhood websites, particularly unofficial ones, and their influence on the organisation, refer to the researcher's article *'Mahdi 'Akif, the Era of Free Expression!'* The article may be found in the book titled: *The Transformation of the Muslim Brotherhood: Disintegration of Ideology and the End of the Organization*, Madbouli, Cairo, 2006.
10. Information on the vision of the Brotherhood and the management of its media campaign was collected by the researcher through interviews with some of the Alexandria Media Committee and media officials in the Brotherhood. All requested anonymity either because of the illegality of the organization and the incurred security scrutiny, or because the publication of some information was not authorised by the leadership.
11. It is the second most important district after the capital, Cairo, in population, politics and economy, and the country's first port. It was founded in 333 BC by Alexander the Great, giving it its name. Alexandria is the oldest and most famous city to carry the name. The Pharos of Alexandria was one of the Seven Wonders of the World, and the ancient Alexandria library was the oldest and most important university of ancient times. It was rehabilitated in 2001. Alexandria was the cosmopolitan city of choice in Egypt until the July 1952 Revolution and the rise in nationalism and nationalisation caused thousands of expatriates to leave the city.
12. The Brotherhood in Alexandria is historically distinguished among Brotherhood organisations in other districts. It is renowned for its initiatives; establishing the first Arts Committee and Political Committee within the Brotherhood. It also spearheaded the candidacy of the first woman from the Brotherhood, Mrs Jihane Al-Halfawi, to run in the parliamentary elections of 2000.
13. A tactic followed by the Brotherhood in every election in which it runs. The names of some candidates are not announced and are kept secret, forestalling any circumstances or developments, such as arrests, that may sabotage their chances to continue their participation.
14. Data and statistics herein are mainly based on the contributions of an official in the Brotherhood Media Committee in Alexandria, as well as papers and documents pertaining to the organisation and which were made available to the researcher.
15. Alawsat, Asharq, 'Muslim Brotherhood Overtakes the Egyptian Parliamentary Elections by Good Methods', *The Middle-East*, no. 9875, 2005.

    For more information on the Sama radio venture refer to: Unrestricted Radio, *Al-Mujtamaa Islamic Magazine*, a Kuwaiti Brotherhood mouthpiece, issue no 1707, 24 June 2006.

    For more information on the importance of the Brotherhood's media performance refer to: 'Lessons Learned', *Al-Ahram* weekly, issue no. 771, 1-7 December 2005

         http://weekly.ahram.org.eg/2005/771/eg8.htm

       Al-Jazeera news network likewise broadcast a report on Sama radio. The Brotherhood's media performance was also tackled by Al-Arabiya news network and BBC Arabic.

16. The researcher discussed this issue in a paper titled: 'Islamists: Dancing to the Beat of Globalization' at the International Conference organised by L'Institut Français du Proche-Orient, Amman, December 2005.

17. For more on the administrative revolution within the Muslim Brotherhood and its impacts on the organisation refer to: *Mouvements* magazine, special issue *'Islam and Modernity'*, Paris, November 2004 *(Sous les éclats du sourire des winners pieux ... 'l'islam par projets', une utopie en mode mineur)* (in collaboration with Patrick Haenni).

PART THREE

Public Opinion

ERIK NEVEU

Measuring and Comparing Opinions: A Practical and Theoretical Challenge

From Magreb and France to Mashrek and Germany, we share thoughts, reflections and debate on public opinion in Europe and the Middle East. As the notion of 'public opinion' is one of the most commonly used by social scientists, politicians and journalists, its meaning can seem obvious. However, for this very reason, it is worth exploring and clarifying the exact meaning of this overused and underdefined phrase.

A first approach could suggest that the definition of public opinion can be related to 'mapping'. In each and every society, people have beliefs and opinions about life in that society – the events and challenges they face, the existence of threats and the behaviour of rulers. According to this minimal 'soft' approach, public opinion is a panoptic representation of the judgements and concerns existing in a population. The notion has a cartographic dimension, as it implies the aim of producing the clearest possible overview of the representatives of a society.

However, if left merely as a general notion of public opinion, the explanation would remain poor and ill-defined. Since its appearance as an element of political vocabulary and a mobilising myth of democratic societies, 'public opinion' has a much more precise meaning. Public opinion has been institutionalised. This institutionalisation is threefold.

The modern notion of public opinion includes a question of legitimacy. Members of the polity are citizens, not subjects. They have rights, and among them is the right to publicly express judgements on social and political issues. To quote the French historian Mona Ozouf, 'The appearance of public opinion as the supreme authority will indicate that the world has been emptied of other authorities. After all, how could what is hailed as queen of the world exist under the rule of kings?'

Public opinion cannot fully develop without a complex machinery of institutions and processes that Habermas links to the notion of public sphere. Public opinion is the product of a specific process of socialisation producing educated individuals able to argue and debate. It also requires the presence and input of media, representative institutions and spaces where people can freely debate, express and compare their opinions.

This institutionalisation incorporates a twofold process of selection. All ventures, all groups and individuals do not have equal opportunities, even in the most democratic of societies, to fuel and influence 'public opinion' and its debates. To paraphrase Orwell, 'All are equal but some are more equal than others', in resources and in media access. At the same time, all questions are not considered as equally worthy of attention. 'Public opinion' is about values requiring action from leaders, about them triggering opposition – thanks to agreement on their 'sacred' status – in society.

This brief and unsatisfactory effort at clarification immediately draws attention to two conclusions. The first recognises how difficult, even impossible, it would be to make a fair and unbiased comparison between two public opinions from different institutional backgrounds and processes. One cannot make a comparison between, for example, the nature of the public sphere in Saudi Arabia and that in a European country, without recognising that they are structured very differently, with a smaller social perimeter in Saudi Arabia than in countries such as France or Britain. The second conclusion is clearly that public opinion is a social construction. A social construction is not an illusion, nor an arbitrary product. The term suggests that both the objective institutions and the subjective beliefs linked to the idea of 'public opinion' are the products of history and social structures and that competing visions of public opinion usually exist in a polity. For example, in her classical study, Susan Herbst[1] shows for instance that staffers in US state parliaments consider the lobbies as the best voices for public opinion, whereas party activists place greater value on opinion polls and journalists than on the conversations they have with colleagues and those they overhear in public spaces.

In a nutshell, public opinion never exists as an objective and unquestionable reality. It always depends on the mediations and tools used to define, measure and mobilise it in narratives. There is not one single universally scientific method but instead there exist a number of methods of analysis, more or less effective in different situations, to produce 'snapshots' however blurred and incomplete. Critically, public opinion remains so mobile, complex and fluid as to continually escape pure objectification.[2] This plea for a constructionist view should not open the door to boundless relativism, or to nihilist criticism. It is simply an invitation to subject any potential process of comparing public opinion to two preliminary questions: which opinions in which societies? And how can one measure them?

Four basic questions on the comparison of public opinion

In an attempt, however over-simplified, to identify some basic questions concerning the social basis of public opinion and the scientific problems that they pose, four parameters will explored.

The first question can be posed thusly: How is society divided? Sociologists have always focused on social stratification. Durkheim's (*De la division du travail social*, 1893) analysis of the social division of labour is well known and Norbert Elias (*La civilisation des moeurs*, 1974; *La Dynamique de l'Occident*, 1975) highlighted the link between the progression of social interdependences and the process of civilisation. One must focus on this issue in order to properly interpret public opinion, including both its creation and its circulation. As an example of the physical factors affecting both these issues one can point to variables such as population density, spatial organisation of settlements, housing conditions, opportunities for spatial mobility and the existence of diasporas and immigrants.

All of these factors have direct and significant effects on the material circumstances allowing (or inhibiting) exchanges and connections between individuals and therefore the circulation of opinion. The range of contacts and the variety of social experiences shape the array of opinions. When debating media and public opinion, one should link the modern notion of mass communication to the more traditional dimension of physical communication and the role of social networks in the production of public opinion. A society divided by a limited number of rival communities may tend towards radical oppositions of opinion – an 'us and them' mentality. Conversely, a complex social division of labour in a society where individuals and groups are conscious of what they owe to society, even to subordinates and rivals, can increase the chances of a 'peaceful confrontation' of opinions. The term 'globalisation', however overused, is useful in this context to explain this phenomenon. Globalisation can be associated with complex and widespread chains of interdependence. Events in China, for example, could mean the closure of a factory in Tunisia. Decisions taken at the Tokyo stock exchange or by the OPEC could have a huge impact in Turkey or Belgium. Such a world is perceived as threatening and incomprehensible by many ordinary citizens and its added complexity often fuels feelings of powerlessness and fear, and can boost the rise of nationalist, fundamentalist and Manichean sentiments.

To what practical conclusions can such theoretical remarks bring us? In fact, they can bring us to very practical ones, such as the need to focus on the different degrees of 'nationalisation' of public opinions in different countries. Is the population of a state sharing a common language or using several different ones? Does the community share a common media-literacy? Or does it share aspects of the same quality and quantity of media consumption, for example newspapers, television and television news programmes? Is the country divided by fragmented 'mediascapes'? Do citizens have equal or unequal access to the media supply? Taking factors such as mediascapes, social stratification, autarky and cosmopolitanism into consideration will help to prevent an essentialist vision of public opinion. It prevents simply relying on the influence of 'culture' or 'civilisation' as an explanation and helps to make sense of the construction of public opinion in terms of and arising from social morphology.

A second question linked to the social basis of public opinion can be expressed as: Who? Who is allowed to have and express opinions? The answer to this comes from law, politics and also personal belief concerning which opinions are worth listening to and which are not. Three brief illustrations can highlight this question. The first issue concerns foreigners. Does a foreigner's opinion matter in the country where he or she is an immigrant? The most common response to this question is no. However, its meaning can be very different. In some countries, such as some Emirates, foreigners make up the majority of the working population. Conversely, the label of foreigner can sometimes be used to discredit some opinions. As an illustration of this, the violent mobilisation in French suburbs in 2006 was often explained by the action of 'young immigrants'. Despite this title, most of the youngsters carried a French ID card in their wallets.[3]

The second issue regards the possession of full citizenship as the 'key' to the right to express oneself and have an opinion. However, one must ask if citizenship is universally defined under the same conditions at, for example, the same age throughout the world? How can we make sense, for example, of the case of the Palestinian *chebab* of the first Intifada, of the paradoxical situation of these young men? They are too young to vote, too young to be counted in an opinion poll; however, their high risk commitment to the Palestinian cause can truly express a Palestinian public opinion. In addition to age, gender must also be considered. Are opinions of groups or individuals to be valued equally, regardless of sex? Such questions are particularly pertinent in Islamic communities. But beyond official principles, its practical answer is not always absolutely clear even in European politics.[4] The symbolic status achieved by working-class mobilisation and opinion is also worth considering. An example of this can be seen in Europe when one considers the extensive overuse of the notion of 'populism'. This notion had the potential to be useful when applied to North or South American historical cases, and when applied to precise movements and ideologies rather than used as a general term. However, as it is, it has become a conceptual sponge used to absorb a wide variety of characters and causes, which has inevitably hollowed its theoretical meaning. But if the word has lost a degree of its meaning or impact, it still retains a function, however changed this may be. As Annie Collovald showed, the word populism has shifted from the identification of an ideology or movement blamed for manipulating the working classes' fears and hopes to being a word which is used to blame working-class beliefs for being the major threat to the democratic public sphere. Today, a large proportion of opinions recorded by polls or votes are described as 'populist'. In the academic world, such a label reinforces a great divide seen between a true 'opinion' – rational and well informed – and working-class reactions and beliefs expressing passion, fear and ignorance, not to be taken seriously. Such a description sounds like a repeat of the nineteenth-century pleas for a vote based on poll tax, as the only guaranteed way to sweep from the circle of respectable opinions most of those coming from the working classes.

The practical difficulties faced when putting these considerations into practice are substantial. How should one go about comparing 'public opinions' made up of different proportions of gender and age? How should one compare public opinions in a situation where all individual opinions do not carry the same weight or value? And when even the parameters of these inequalities are different from one country to another?

The social structures that shape public opinion raise a third question to be asked when making comparisons, which can be posed thus: What are public opinions about? It is always possible to gather opinions on a wide range of issues, from views on sex appeal to favourite meals. Most of the time, opinions can be described as 'public problems': the only ones that become objects of public policy and public debate are those identified (often after time, struggles and demonstrations) as requiring discussion, possibly followed by action. It is an established fact that not every question debated in day-to-day life has equal opportunity to give birth to sentences starting with 'Public opinion says that …' For example, issues such as war and taxation are much more likely to become a matter of public opinion than an issue regarding topics such as sex, food or dress. Even when sex does become a matter of public opinion, it is almost always presented as a question of morality, freedom and normality – virtually never as a question of pleasure or aesthetics. To gain enough attention to reach the status of 'public opinion', subjects need to be linked to questions such as morality, national identity, gender patterns or health and safety. Each society 'censors' itself by defining, according to its history, culture and traditions, which topics become objects of public debate, and which remain 'taboo' and locked in private discussions. In French the verb *opiner* means both to produce an opinion and to agree by a nod of the head. This is a perfect example of the way that every society has its *doxa*; in other words, issues on which one is supposed to agree – it is, in a sense, taken for granted. However, the private, the public and the 'taken for granted' rarely overlap from one society to another. One of the effects of globalisation of the media is to publicise these differences between countries around the world, allowing, for example, the inhabitants of country B to see what is at stake

Measuring and Comparing Opinions: A Practical and Theoretical Challenge

as a result of public debate in country A. The affair of the caricatures of the prophet Mohammed – explored in this book – as well as the campaign against Salman Rushdie's *Satanic Verses* are clear examples of what can arise as a result of this situation. In both these cases, the ensuing scandal was triggered by the fact that something taken for granted by the majority in country A was challenging the beliefs of the majority in country B. In this case, the issues at stake were the holiness of the prophet and the freedom of the writer. One of the practical consequences of these collisions of public spheres, cultures and national opinions may be the dangerous process of cross-radicalisation of prejudices. European intellectuals are often highly critical of the representation of Western or Jewish culture in the Arabic media, sometimes with good reason.

However, can we really be prepared to accept the loose and even ridiculous overuse of the term 'Islamism', in even the most serious and well-respected Western media, to refer to anyone from al-Qa'ida to the pilgrims of Mashad, the Turkish government and or any bearded *hajji*?

A fourth and final question can also be asked regarding the process of establishment and circulation of public opinion: Which are the institutions allowing or producing a public opinion? The masterpiece of Jurgen Habermas on the 'Public Sphere' can be challenged and discussed;[5] however, it still offers a stimulating analytical framework. It asks how private relations and family networks can boost, or damage, the ability to produce and express critical opinions. It describes how a network of spaces and institutions for discussion and communication, including salons, coffeehouses and Masonic lodges in eighteenth-century Europe, processed these private opinions first into semi-public debate on cultural and literary issue, and then finally into full 'public opinion' debating political issues. However, one of the limits of this conceptual masterpiece can be easily identified: Habermas is European. Political scientists or media scholars are usually not great readers of regional studies and they may be blind to the nature of institutions used in different societies to transfer private opinions into the public sphere. When studying the rise and usage of the right to petition in seventeenth-century England, the American historian David Zaret[6] expressed a thought-provoking idea. He suggests that to fully understand the process of public opinion, researchers should stop using terms with capital letters – such as Public Sphere, Religion or Class – or more precisely, only use them after studying in depth the practical details of everyday communication and experiences. How can we understand public opinion in the Middle East if we do not consider the details of day-to-day social networks and structures of communication? Religion is not purely a question of faith. It should be considered in terms of structuring social relationships, including pilgrimages, discussions, sermons and the use of the mosque for living and socialising. Shopping in Alep's bazaar or on the streets of old Cairo is not the same experience as going to the supermarket. In addition, without adequate knowledge of the culture, how can one pay the appropriate attention to the importance of the tradition of oral communication and the vitality of street life which is so powerful in Iran? The fear of ethnocentrism[7] should not, however, prevent us from questioning the existence of a fully 'public' opinion, in its Habermasian definition, in Arabic and Middle-Eastern countries. The empirical and non-normative answer is that, in many cases, such a thing does not exist. The restrictions placed on a complete, free and public use of reason and speech are important, as many opinions do not have the space, right or opportunities to be heard or read in a public sphere. For at least two reasons, to be discussed later, such an observation should not lead one to the conclusion that comparisons are pointless. The first reason for this comes from the combination of traditional social networks with the use of modern communication tools: internet and strong gender segregation, state control of the press but hidden satellite dishes in the garden that connect to information channels that borders, uniforms and mullahs alone cannot contain. The second is that before and below the institutionalisation of a labelled public opinion, the near silent simmering of resistance and 'hidden transcripts'[8] is an unavoidable stage.

Poor, reflexive use of language brings citizens and researchers to speak of Danish, Lebanese and Saudi Arabian public opinion. If such phrases express the fact that the citizens of these countries have opinions – whether they are whispered, spoken or even printed – or that they share an equal cognitive ability to express judgement on private and public affairs, then they make sense. Usually, however, 'public opinion' carries more and heavier prejudices: the public opinion is or must be unified like the Umma, that the opinions can finally be expressed in figures (N per cent of Emiratis, y per sent of Swiss believe that ...), that such figures express actually comparable things. The caveat suggested here as a provisional conclusion is both modest and devastating. Intellectuals and stakeholders should never forget that the same phrase – public opinion – covers conditions of production, networks of communication, definitions of those who can legitimately produce an opinion and perceptions of the topics on which disagreeing is accepted, which are very different. Public opinion is not enough to allow us to give up the idea of measuring and comparing. It is more than enough to prompt us to automatically measure and compare.

Mapping and measuring public opinions

How can public opinion be perceived and measured? The answer seems crystal clear for mainstream and central-Western social science: by using opinion polls! Opinion polls have the potential to be precious tools of evaluation for the state of public opinion. However, three reasons encourage us to realise the limits of such a technique. The first one can be linked to the 'modern' legitimacy of this technique. To put it in a nutshell, using opinion polls to establish public opinion is a modern construction with barely fifty years of experience.[9] But public opinion has always belonged to political language, and had influence even before Gallup. Therefore, one should pay attention to the different historical forms of opinion. A second reason indicating the limits of relying on opinion polls is that it is not always possible to just ask a simple question due to the conditions under which the inquiry has to be organised, as can be seen in Baghdad or Kerbala today. Finally, one should also keep in mind the limits and blind spots of such a method. For example, opinion polls were far from predicting the success of Le Pen in the second round of the French presidential election in 2002 and they have rarely predicted the great revolts, riots and protests that have shaken societies throughout history, from the European '68 upheaval to the Iranian Revolution. These are the reasons for which alternative, more qualitative methods of measurement are necessary. They can be called symptomatic methods – in other words, they are based on the study of behaviour of a society and the signs reflecting its deepest feelings and beliefs.

Statistical methods

The statistical method of analysis, via the use of opinion polls, is well known and often used. The process is simple, at least to summarise – one selects a sample big enough to limit the margin for error, and ensures that it is close enough to the actual structure of the population studied to be representative of it. Participants are then surveyed in order to reach a conclusion. However, this method of analysis is subject to two problems.

The first can be described as technical as it concerns the structure of the sample. Defining an appropriate sample requires an in-depth knowledge of the population being investigated (the '*population-mere*'). The 'population' can refer either to the nation or to the people living within the boundaries of the state – this is a critical difference. This knowledge of the population is usually produced by a network of state institutions, which collect and treat data such as population census information, employment statistics and immigration figures. However, as highlighted by Alan Desrosieres in his comparative history of state apparatus for data collecting and processing,[10] such a result requires a combination of investigative capabilities and standardisation of gathering techniques which can only work with the mobilisation of financial resources, specialised bureaucracies and a minimum level of trust in the state by the population. The accurate production of such data cannot be the result of a close watch over a population by, for example, specialised surveillance forces or secret police. Nor can it be the result of heavily biased or censored reports, designed with political means in order to frighten or flatter a ruler or *ra'is*. Even the most carefully produced statistics on population are the result of struggles. For example, the production of data related to ethnic identity is a topic of fierce and ongoing debate in France. Equally, measuring the weight of different religious communities is a sensitive issue in Lebanon. In addition, once a sample is selected, another technical question arises regarding the right population quotas. The percentage of US citizens refusing to answer public opinion polls is now close to 25 per cent. Similarly, any French pollster would describe the difficulties in reaching the necessary quotas, especially in certain working-class, less-educated population groups, and how such a difficult situation can lead to inaccurate methodological choices.[11] Finally, questions must be fair and answers clear, and both must be unbiased. Opinion survey companies are skilled to face these challenges – a fact which does not always prevent the presence of inappropriate and difficult questions, or the over-interpretation of some answers.

This last point leads to a more epistemological criticism, developed by Pierre Bourdieu in the title statement of his famous paper, 'Public Opinion does not Exist'.[12] When surveying public opinion and transforming answers into percentages, researchers often make two common mistakes. The first fallacy is to assume that everyone has an opinion. This is not the case, especially when surveying opinion on difficult subjects such as foreign affairs. Questions can be very far from the practical experience of the majority of citizens and such questions lead to low levels of response. One of the skills of a good pollster is to prevent silence and 'non-answers' and produce the maximum response possible. When football was less popular than it is now, a French political scientist[13] had the funny idea of having a questionnaire concerning the French football team published in an academic journal. Such questions were as meaningless then for a majority of the academic readers as those on Cyprus' reunification would be for many retired farmers. The second fallacy is the belief that all opinions are equal. In theory, all do have the same value and dignity. However, they are not all the result of the same process. Some are produced using information and criteria that fit the question. Others simply mirror class ethos or misunderstandings. During the mid-nineties, a French research on social policies[14] showed that when people who answered yes to the question of whether or not to reduce welfare expenses were invited to explain which expenses they were willing to reduce, their answers sometimes targeted public health expenses. More often, however, they quoted the army budget and wages of French MPs. In an ordinary opinion poll, which would not have any 'open' questions of this nature, the answers to this question would simply have been combined, leading to an (inaccurate) conclusion suggesting that a majority of citizens supported the reduction of the welfare state.

Finally, it is important to notice that opinion polls are costly and, as the old adage goes, 'he who pays the fiddler chooses the tune'. This does not mean that the results of studies are pre-defined by their benefactors or financial backers, but it does mean that the questions 'worth asking' are asked. The addiction of the French media to opinion polls produces weekly rankings of the popularity of politicians, although it would be hard to find opinion poll data that predicted the violence in the French suburbs in 2005.

The symptomatic tools

The symptomatic tools are based on the interpretation of signs and behaviour considered as a reflection of public opinion. Naturally, they are usually more complicated to translate into measurements or ratings. Remember that the story of public opinion has, until very recently, been based on the use of these symptomatic tools.[15] In an over-simplified approach, four types of these symptomatic tools can be mentioned.

The first can be linked to the idea of an *opinion survey*, for example, of a watchdog observation of the speech and behaviour of authorities or their agents. In this, we are between repressive supervision and reports on the state of public spirit. Civil servants, often belonging to police bodies, behave as the eyes and ears of the authority. French historian Arlette Farge describes policemen of the *ancien regime* who are referred to as flies-on-the-wall (*mouches*), who would listen to talk and rumours in cafes and other public spaces. In contemporary France, a euphemistically named department *Renseignements généraux* or 'general information' still collects data and produces reports on the state of public opinion. One must remember that even the *The One Thousand and One Nights* are filled with stories of *khalifes* hidden under ordinary clothes who patrol the bazaars, listening to chat and gossip in order to check the popularity of the *wazir*. In fact, the Jordanian press reported similar behaviour from the young King Abdallah following his accession to the throne. There are always risks of fallacies in this area, whether aiming to produce anaesthetic reports to flatter rulers or from over-subservient special forces hoping to improve the police budget.

A second symptomatic tool can be compared to the *representative government applied to opinion*. One can define public opinion as the global map of the beliefs and judgements coming from elites and speakers considered as the 'voices' of society. Patrick Champagne has shown how, in France, public opinion meant firstly the opinion of an enlightened elite – writers, elected authorities and members of parliament. This pattern of an elite-driven opinion will later be democratised with the role given to the press. The morning press will come to be seen as a reasonable and accurate portrayal of public opinion. It is unnecessary to point out that such a delegation of representation is also vastly ambiguous. How can one be sure of a true correspondence between journalistic opinion and that of the public? How can the elite know the depths and variations of public opinion?

A third level of symptomatic tools can be linked to the idea of *mobilised opinion*. In this case, citizens behave proactively in the public sphere. They transform opinions into voices, using what Charles Tilly called 'collective action repertoires'. Citizens sign petitions, go on strike, organise demonstrations in the street or react in a certain way to public appearances of the ruler – do they applaud, remain silent, or shout? There exists one common point between these 'mobilised' opinions and statistical opinions: quantification is possible. One can give figures of strikers or demonstrators, and beyond certain thresholds these numbers can indicate, to a certain degree, the state of public opinion, for example in the case of the mass demonstrations in Lebanon after the murder of the prime minister Rafiq Hariri. The sociology of social movements draws attention to the ambiguities of these mobilisations. All groups do not have equal resources in funding, nor do they have equal access to influential media connections. Small, highly publicised and highly funded demonstrations can receive heavy, positive media coverage, whereas others, perhaps reflecting deep social sufferings, may receive hostile media comments and coverage. Even action with an apparently obvious meaning can be misinterpreted and trigger misunderstandings. In the recent book by Nicholas Mariot,[16] he shows that the crowd applauding the French president during his trips and public appearances does not always indicate their popular support and agreement. Regardless of whether the audience is made up of militants or schoolchildren, clapping hands may simply express politeness, or perhaps the excitement of a public gathering, without having any deeper political meaning. Public elation can come from something as simple as an extra day of holiday or perhaps the celebrations attached to a political event. Such a reaction can give rise to a false impression of happiness and helps to explain apparent jubilation visible at each of Saddam Hussein's birthdays.[17] One should finally keep in mind that public opinion mobilised into physical action always has to face the challenge of its presentation and coverage in the media. Media has the potential to act as a loudspeaker to a cause, when it provides sympathetic coverage. However it also has the power to jeopardise the actions of a cause or campaign. A perfect example of this is the portrayal in the French and European press of the build up to the two Gulf wars and the anti-war demonstrations in Arab cities, which were portrayed as fanatical Islamism and the 'hysterical' behaviour of Arab street crowds – '*la rue arabe*'. The behaviour of the media reminded one of the nineteenth-century anti-democratic paintings of Gustave le Bon,[18] who portrayed hysterical, dangerous, 'female' crowds running riot.

Finally, a fourth aspect of *symptomatic tools* should be considered. One must ask what happens when and where there is no freedom and no outlet for the expression of opinion. Should one immediately conclude that the absence of freedom and the prohibition of public debate smothers public opinion? Or does it merely smother the expression of public opinion? Non-democratic regimes do not allow a public sphere (defined as a network of institutions and groups channelling an open, accessible and competitive discussion on public affairs). They challenge the very right and opportunity to have and express a genuinely 'public' public opinion. However, they can never entirely suppress the ability of the citizens to have opinions and feelings regarding the social organisation. One needs to pay attention here to what James C. Scott labelled 'the arts of resistance' or what Jean-Francois Bayart labelled '*bottom-up politics*' or 'politics of the belly'.[19] Research into undemocratic regimes must investigate behaviour that lacks immediate or obvious political meaning. It must decipher the signs and the symptoms of attitudes concerning power and politics. The use of humour, passive resistance, abuse or exploitation of rituals or traditions (such as religious processions) or even destructive or predatory behaviour against public property all become ways of expressing and sharing opinion. Literature can help shed light on this issue. The Egyptian novelist Albert Cossery describes the possible political impacts of laziness or mockery. In Naguib Mahfouz's masterpiece trilogy, one should also remember how the old al-Gawaz instructs his sons not to speak to British colonial soldiers living in their street, as a protest against their presence. Ethnographers also offer precious materials regarding these hidden expressions of opinions, which Asef Bayat emphasises heavily when he describes the political potential of the 'quiet encroachment of the ordinary' by street merchants and illegal constructions in the public space of Tehran. The comprehensive, almost ethnographical reporting of the Israeli journalist Amira Has on ordinary life in Gaza helps us more than many opinion polls and surveys to make sense of certain areas of Palestinian opinion.

Once again, one must mention the difficulties regarding *measurement*. Problems can arise when deciphering behaviour which in fact has the opposite meaning from that which seems apparent at first. Gilles Kepel[20] shows that in Egypt calling even the lowliest civil servant *bey* or *pacha* or replying with *hadir* to a question or sentence can be very ironic, and the *pacha* can become *abd-el-rutin* (slave of routine) as soon as he turns his back. Consequently, an observer can over-interpret or misinterpret these micro-events. As for passive resistance, it can be the hidden expression of repressed opinion, but can also be seen as a safety exit accepted by rulers for powerless people, or even as a simple expression of selfish behaviour.

This brief overview has succeeded in producing more questions and caveats than firm conclusions and solutions. But caution is vital when discussing and investigating the notion that Bourdieu describes as 'the equivalent of God' in democratic beliefs, and the first rule of the comparative approach is to establish exactly what is to be compared, and what is comparable.

Focusing on the nature, the qualitative dimensions of the making of public opinion or its symbolic status, might tend to suggest – with a touch of evolutionism – that European societies are 'more advanced' than those of the Middle East. The processes of institutionalisation and legitimisation are more developed in Europe. The networks of social division of labour are long and wide enough and the resource inequalities between groups limited enough to leave space for non-violent conflict between competing political visions. One should emphasise that the idea of a European public sphere is more a mirage than a shared experience or reality. There also exist two possible peculiarities of Arabic or Middle Eastern opinion. The first is that the existence of the Arabic language implies a much wider potential for communication than that of the use of English in Europe. Simultaneously, for reasons highlighted by Olfa Lamloun regarding al-Jazeera, transnational media exists from which the message is not necessarily the docile celebration of all regional rulers. The challenge for lay citizens is to connect themselves to this media, whether through legal or illegal means, and this belongs to the 'art of resistance'. One may thus see in the realm of the public sphere what Gerschenkron analysed as the paradoxical advantage of economic backwardness: the opportunity to 'leapfrog' over certain stages of development, which is far faster than following each stage of the process.

If social scientists are allowed to express wishes, let's hope that the combination of political mobilisation, improved education and technological resources will allow us to debate this 'great leap forward' of the mosaic of Middle Eastern public spheres.

Notes

1. Herbst, R., *Reading public opinion. How political actors see the political process*, Chicago: Chicago University Press, 1998.
2. The remarks developed by Jocelyn A. Hollander in a reflexive contribution on the limits of focus groups could be applied to any methodology of approach of opinion (individual, group, public). 'Research participants are not uncomplicated information storage facilities who need only the proper instructions to open their hearts and minds to the researcher. Instead they are complex, often contradictory mosaics of history, experience, motivation and interests. Focus groups like other methods, provide one window on these mosaics. The participants in focus groups are not independent of each other, and the data collected from one participant cannot be considered separate from the social context in which it was collected.' (2001; 631).
3. The French language uses terms such as 'second generation immigrants' and 'young people from the immigration' as if being an immigrant was a hereditary disease.
4. Achin, C. and Levêque, S., *Femmes en politique*, Paris: La découverte, 2006.
5. Calhoun, C., (ed.), *Habermas and the Public Sphere*, The MIT Press, 1992, and François, B. and Neveu, E., *Espaces Publics Mosaïques*, Rennes: PUR, 1999.
6. Zaret, D., 'Petitions and the 'invention' of public opinion during the English revolution', *American Journal of Sociology*, no. 101 (6), 1996, pp. 1497–1555.
7. Here the risk of ethnocentrism is doubled. On the one hand ethnocentrism could over-emphasise the 'traditional' dimensions of Middle-East societies – attributing to them the irresistible influence of religion, oral culture and traditions, forgetting that these dimensions are embedded in the modernity of electronic media, diasporic connections. On the other hand ethnocentrism could lead to too much emphasis being placed on the rational, deliberative and public functioning of the public sphere in the more developed countries. In a stimulating ethnography of a village in the south of France, Phillipe Aldrin (2003) shows the remaining strength of street culture and family connections in the relationship between local opinion and politics in contemporary France.

8. Scott, J., *Hidden Transcripts*, Yale University Press, 1990.
9. Champagne, P., *Faire l'opinion*, Paris: Minuit, 1989; Blondiaux, L., *La fabrique de l'opinion. Une histoire sociale des sondages*, Paris: Seuil, 1998.
10. Desrosières, A., *La politique des grands nombres. Histoire de la raison statistique*, Paris: La Découverte, 1993.
11. Such as multiplying by two or three the weight of answers coming from very small samples among unreachable demographic groups of voters, which leads to the risk of increasing margins of error.
12. Bourdieu, P., 'L'opinion publique n'existe pas', in *Questions de sociologie*, Paris: Minuit, 1980, pp. 222–235. See also Lehingue, P., *Subunda*, Bellegarde: Ed du Croquant, 2007.
13. Offerlé, M., 'Dépaysement : le nombre des buts', *Politix*, 1989, no. 2 (7), pp. 151–154.
14. Gaxie, D., 'Au delà des apparences', *Actes de la Recherche en Sciences Sociales*, no. 81–82, 1990, pp. 97–113.
15. Champagne, P., *Faire l'opinion*, Paris: Minuit, 1989; Farge, A., *Dire et Mal-Dire, L'opinion publique au XVIII°*, Paris: Seuil, 1992.
16. Mariot, N., *Bains de foule*, Belin, Paris, 2006.
17. Baran, D, *Vivre la tyrannie et lui survivre: L'Irak en transition*, Paris: Mille et une nuits, 2004, pp. 78-91.
18. One of the most interesting results of our conference has been to establish the differences in meaning, regarding the notion of the 'Rue Arabe/Arab Street' in Europe and the Middle East. Whereas the phrases suggested for our Arabic colleagues the existence and vitality of a loosely structured public opinion even in non-democratic Arab countries, its dominant meaning in Europe is much more deprecating. It suggests over-emotional and non-rational opinion.
19. See also a recent overview of these approaches developed by Olivier Fillieule and Mounia Bennami-Chraibi (2003; 43–126).
20. Kepel, G., *Le prophète et le pharaon. Les mouvements islamistes dans l'Egypte contemporaine*, Paris: La découverte, 1984, pp. 222-223.

Bibliography

Achin, C. et al ii, *Sexes, Genre et politique,* Paris: Economica, 2007.
Achin, C. and Levêque, S., *Femmes en politique*, Paris: La découverte, 2006.
Aldrin, P., 'S'accomoder du politique. Economie et pratiques de l'information politique', *Politix*, no. 64 (16), 2003, pp. 177–203.
Baran, D, *Vivre la tyrannie et lui survivre: L'Irak en transition*, Paris: Mille et une nuits, 2004.
Bayart, J-F., *L'Etat en afrique. La politique du ventre,* Paris: Fayard, 2006.

Bayat, A., *Street Politics, Poor People's Movements in Iran*, New York: Columbia University Press, 1997.

Blondiaux, L., *La fabrique de l'opinion. Une histoire sociale des sondages*, Paris: Seuil, 1998.

Bourdieu, P., 'L'opinion publique n'existe pas', in *Questions de sociologie*, Paris: Minuit, 1980.

Calhoun, C., (ed.), *Habermas and the Public Sphere*, The MIT Press, 1992.

Champagne, P., *Faire l'opinion*, Paris: Minuit, 1989.

Collovald, A., *Le 'populisme' du Front National; Un dangereux contresens; Raisons d'agir*, Bellegarde: Ed du croquant, 2004.

Cossery, A., *Un complot de saltimbanques*, Paris: Joelle Losfeld, 1999.

Cossery, A., *La violence et la dérision*, Paris: Joelle Losfeld, 2000.

Desrosières, A., *La politique des grands nombres. Histoire de la raison statistique*, Paris: La Découverte, 1993.

Farge, A., *Dire et Mal-Dire, L'opinion publique au XVIII°*, Paris: Seuil, 1992.

Fillieule, O. and Bennani-Chraïbi, M., *Résistances et protestations dans les sociétés musulmanes*, Paris: Presses de Sciences Po, 2003.

François, B. and Neveu, E., *Espaces Publics Mosaïques*, Rennes: PUR, 1999.

Gaxie, D., 'Au delà des apparences', *Actes de la Recherche en Sciences Sociales*, no. 81–82, 1990.

Has, A., *Drinking the sea at Gaza. Days and nights in a land under siege*, Metropolitan Books, 1999.

Herbst, R., *Reading public opinion. How political actors see the political process*, Chicago: Chicago University Press, 1998.

Hollander, J., 'The social context of focus groups', *Journal of Contemporary Ethnography*, no. 33 (5), 2004, pp. 602–637.

Kepel, G., *Le prophète et le pharaon. Les mouvements islamistes dans l'Egypte contemporaine*, Paris: La découverte, 1984.

Lasfar, A., *Le monde arabe dans la presse française pendant la 'guerre du Golfe' (août 1990–février 1991). Contribution à l'analyse des catégories de l'entendement journalistique*, PhD, Rennes University, 2003.

Lehingue, P., *Subunda*, Bellegarde: Ed du Croquant, 2007.

Mahfouz, N., *Impasse des deux palais, Le palais du désir, Le jardin du passé*, Paris: Le livre de poche, 1989.

Mariot, N., *Bains de foule*, Belin, Paris, 2006.

Offerlé, M., 'Dépaysement : le nombre des buts', *Politix*, 1989, no. 2 (7).

Ozouf, M., 'Quelques remarques sur la notion d'opinion publique au XVIII° siècle', *Réseaux*, no. 22, 1987, pp. 79–92.

Scott, J., *Weapons of the Weak*, Yale University Press, 1985.

Scott, J., *Hidden Transcripts*, Yale University Press, 1990.

Tilly, C. *The Contentious French*, Harvard University Press, 1986.

Zaret, D., 'Petitions and the 'invention' of public opinion during the English revolution', *American Journal of Sociology*, no. 101 (6), 1996.

KAI HAFEZ

European-Middle Eastern Relations in the Media Age

It is not easy to find a theoretical basis for explaining how international relations are influenced by mass media. Traditionally, realism in international relations thinking saw relations between states as being determined by the interests of national governments, but not by those of the mass media. Liberalists, globalists, postmodernists – they all believe that the media are important vehicles for all kinds of transnationalisation processes, but they hesitate to consider the media as core actors of international relations. As I see it, they are quite right to do so, because, despite all the ongoing debates about globalisation, about connectivity, the 'media age' or even 'media democracies', it is my understanding of the current state of research that we have not enough evidence that the traditional players – mostly governments – have been pushed aside by the media or that it is corporate transnational media that rule world politics.[1] The mass media can certainly influence international relations but transnational media connectivity is still too underdeveloped to escape the logic of the nation state and to create global interdependence.

This chapter is not construed as a whodunnit criminal story; the conclusion is not saved until the very end. I do not believe that the over a thousand-year-old dichotomy of Orient and Occident and of Islam versus the West is currently being transformed into a 'global public sphere' with politics and societies at both ends using the media as a vehicle for debate about political and social issues and problems that they would like to solve in tandem. Is Europe evolving as a kind of bridge between the United States and the Near and Middle East? I doubt it. Even though I have always thought that Samuel Huntington was wrong to speak of deeply and essentially antagonistic 'civilisations' in the West and Islam,[2] I am afraid that there is a real danger that nationalist and religiously fundamentalist hegemonies in the world's media systems could merge into cultural warfare and could help to create conflicting ideologies of the 'West versus the Rest'. Huntington's notion of the clash of ancient civilisations is wrong because he failed to understand that it is not the clash of civilisations but a lack of communication that is at work. 'Why do they hate us?' Well, if they do, I guess it is because somebody told them to, and it is television and radio more than books which reach the masses.

Media coverage and intertextuality in Euro-Middle Eastern relations

Let's analyse the problem step by step. The basic question is: what kind of judgement can be made about the quantity and quality of the content of foreign reporting in the West and in Europe about the Middle East – and vice versa? Does the condition that it is in make political dialogue viable? Can we talk of 'intertextuality' in the sense that the argument and the interest of 'the other' is present in Western and in Middle Eastern media?

The answer is not easy but somewhat alarming. Since we have representative content studies of Western media coverage about the Middle East, we are in a position to state that our image of that region is at best fragmented and rudimentary. There is hardly any disagreement among scholars that the Western mainstream mass media's agenda focuses on very narrow aspects of life in the Middle East. Let me give you a few examples.

The quantity of Middle East reporting in countries like Germany has grown in the press over the last decades, but not on TV. The consumer is exposed to an average of two to three articles about the Middle East in a national newspaper per day, which is, of course, much less information than he or she receives about domestic affairs. TV news and reporting are usually extremely event-centred, with terrorism and wars being covered at length, but in an extremely discontinuous way. Neither 9/11 nor the Iraq war have led to a permanent growth in Western television reporting about the Middle East. On the contrary, Pippa Norris[3] from Harvard argues that audience interest has decreased since the end of the East-West-conflict. Since I doubt that, due the constraints of everyday life, more news about the Middle East could be digested by Western audiences, I am not arguing in favour of more news. Nevertheless one could plead for more contextualised and continuous news and information.

About four fifths of the German newspaper and magazine reports over the last thirty years have dealt with political issues. Economic information is below 10 per cent, and other subject areas like culture, entertainment, tourism and religion (in the narrow sense of religious teachings and practices) do not make up more than perhaps 2 per cent of the coverage.[4] As a result, the diversity of everyday life escapes media attention. The Middle East is a zone of political and sometimes economic interest to Western media – nothing more. Interestingly enough, there were decades after the Second World War when entertainment and cultural aspects of the Middle East were front-page news in the West; for example, stories about the Persian Shah and his wife Farah Diba made front-page news even in magazines like *Der Spiegel*. Looking at the statistical news development, it is quite obvious that the Six-Day War of 1967, in particular, was the first of a series of political shockwaves that went through the Western media, politicising the news and changing the whole fabric of news-making. The second shock that transformed what communication scientists call 'news values' – the principles guiding news selection – was the Iranian revolution in 1978–79. It is much too simplistic to hold that Western culture is inherently based on stereotypes of the Middle East. The change in news values over decades makes it much more plausible that great events tend to create their own news standards. In the case of Western reporting on the Middle East, it has been a step-by-step worsening of standards and the creation of so-called 'news routines' in the selection of topics that all players – news agencies, newsrooms and audiences – adhere to. It is about time to question this primacy of political news-making and come up with a clearer view of Middle East life and culture.

The existing content analyses show that there is an extensive concentration on conflicts and violence in our Middle East coverage. While violent events no doubt need to be reported, they tend to be over represented. An analysis of some 12,000 articles in German newspapers has revealed that 50 per cent of all articles on Islam are connected to violence[5] – as if that was an authentic view of a religion that 1.2 billion people adhere to worldwide. Coverage of other spheres, in which Islam does not play a dominant role, for example the Palestinian conflict with Israel, is reported in a more 'civilised' way, with much more reporting about regular politics and diplomacy. But news and reports where 'Islam' is involved create the image of a chaotic Middle East. Although media effects on people's opinions are hard to gauge by scientific means, it seems plausible that the concentration on negative news creates the widespread feeling in Western societies that the Middle East is a dangerous place – an area of the world where you'd better not go.

Comparative research in various media systems of the world has shown that such news standards are a universal feature. Asian media tend to report about the West in pretty much the same way as Western media do about the East. Political violence in Ireland or Spain or Islamist terrorism make for preferential news. But there is also a certain north-south gap in the sense that developments in European and US societies get more attention in non-Western media than non-Western societies get in the media of the West, which demonstrates that the USA and Europe are at the centre of the global news system. Despite all recent political conflicts over the last years, culture in the West – not only Europe, but also the USA – has received greater attention in the Middle East than vice versa. The West is 'metropolitan' news in the world's media.

Such disparity of news values shows a deep divide in each society's values on both sides. Where one side is interested in the life and culture of 'the other', 'the other' is not, so 'dialogue' remains but a political catchphrase with no real impact. The terrorists' effort to force attention is not helpful and has made Western societies even more self-centred and eurocentric. However, there is another field where similarities between Western and Middle Eastern media are perhaps more significant than the differences; namely in the way narratives frame the way stories are told and events are interpreted. I am quite aware that in a radical philosophical sense, 'objectivity' does not exist. But there are various truths, and truth is certainly 'what man can't change'. Yesterday's massacre is yesterday's massacre: you might not agree on the causes that led to it, but it should be possible to agree on the number of dead, for instance, as we should be able to agree on the number of people who have been killed in Iraq and who are still being killed. Besides the search for such 'absolute truths', journalism is emboldened to report in a 'balanced' way, which usually means that news is duty-bound to consider the arguments of all the relevant parties within a conflict. In the end, there are ways to conceptualise narratives and the intertextuality in news.

Through having conducted a number of case studies about Western conflict reporting on the Middle East[6] I have to come to the following conclusion. There are cases in which a greater balance of reporting can be witnessed than in other cases, and there are rules as to whether one-sidedness or balance prevails. This can be seen in the coverage of two different conflicts. In Germany and in many Western media, the coverage of the Israeli-Palestinian conflict has undergone tremendous changes over the last decades.[7] While in the sixties and early seventies, support for Israel was often overwhelming, the perception changed in the course of Israel's invasion of Lebanon in 1982, the first intifada of 1987 and the Oslo agreement in 1993, allowing for much more balanced reporting. There is still a notion in the Western media that I would call 'in dubio pro Israel', but things have changed; especially in Europe – much less in the USA – the cause of a Palestinian state is well established. There might be imbalance in the presentation of specific news events, but on the whole, coverage trends to be rather balanced. Protagonists of both governments get ample space to air their views. Atrocities perpetrated by both Israel and Palestine, as well as massacres and terrorism, are covered.

Such analysis is based on long-term observations; so let us look at the short-term case of the early days of the war on Iraq, which, in the narrow sense of a fully-fledged war, lasted for about four weeks in 2003. In the United States during the war, almost all mainstream media, the press as well as radio and TV, supported President Bush or at least did not allow any real criticism. Germany's media were much more pluralist in their approach, allowing anti- as well as pro-war voices to express themselves in articles, talk shows, etc. While 80 per cent of Germans, as well as the government, were against the war, the media were pretty diverse. The situation seemed comparable to countries like Spain, with the difference that the Aznar government was pro-war, although the Spanish people were against it, and the Spanish media system comprised various elements from pro-government TV to critical newspapers.

On the basis of those two cases, and of course there are many more similar cases, my interpretation of Western framing and story-telling points in different directions. Mass media in Western countries that experience existential crises or that engage in full-scale wars seem ready to 'rally round their flags' and support their governments and what they define as their interests. On such occasions the media reproduce the division of hemispheres – of Orient versus Occident – and dialogue in the mass media, where it exists at all, collapses. For short periods, the media's ability to act in a pluralist way and give weight to Middle Eastern perspectives and argumentations can be seriously endangered. This is what Hannah Arendt, the well-known German philosopher, observed during the Vietnam War and what she labelled the 'mentality of raison d'état'[8] – an absolutely serious threat to the plurality in Western democracies. The case of Spain, as a country that was involved in Iraq but still upheld diversity in the media, does not serve to disprove this rule because the military engagement of Spain remained rather limited and did not activate the rally-round-the-flag syndrome. Much more interesting seems the British example. I have done a small content analysis of several British newspapers' coverage of the war and was quite impressed by the relative diversity upheld even during wartime. Although there were patriotic trends, this was certainly a much more distanced coverage than, for instance, during the Falklands War.

The development of reporting on the Israeli-Palestinian conflict shows that apart from times in which a media system's own country goes through crisis or war, German and other Western media are well capable of retaining a critical distance. While mass communication during crises in which one's own country is involved poses a real problem, seen as a whole, Western conflict reporting fares much better. The reporting of the Israeli-Palestinian conflict and the coverage of the Iraq War in 2003 in countries that were not part of the war alliance show that if countries do not have to rally round the flag or if the conflict as such is of a long-term nature, as in the case of the Israeli-Palestinian conflict, the potential of Western media to uphold diverse images of Middle Eastern conflicts is increasing. During the Iraq War, European and the Middle Eastern media were probably closer to each other than European and American. US colonial-style policy was utterly criticised in most European media systems, as it was in the Middle East, and the European and Middle Eastern quest for multilateralism was an expression of shared values between Europe and the Middle East that ridiculed all talk of a 'cultural clash'.

But let us take a closer look at Arab media. While I know of only few solid content analyses of Arab media, I argue that after 9/11, Arab media developed their own qualities as well as biases when covering foreign news, especially concerning regional conflicts. Lawrence Pintak argues that the West and the Arab world, for instance, inhabit separate information ghettos.[9] For years, I was the advisor to the German government's German-Arab dialogue, bringing Arab journalists together, and I recall that already in the late 1990s many Arab journalists, not only American or British, started to approach the new Arab satellite broadcasting media with a mixture of great respect and severe criticism. There is no doubt that transnational Arab television provides the world with images of Palestinian and Iraqi victims that went previously unnoticed by Western media. It is equally obvious that al-Jazeera, of all the Arab news networks, has been the most willing to integrate 'the other opinion'. Israeli, American and many other voices are heard on the network. In contrast, the leading American television network, Fox News, showed no original interviews with Arab politicians during the Iraq War of 2003. Also, I see no sign that mainstream Arab TV is falling into the trap of identifying Europe and the USA as one and creating a coherent view of an enemy called 'the West'. That is the case with many Islamists, but not with mainstream media.

However, critics have often bemoaned the fact that there is a clear pan-Arab bias with regard to the selection and interpretation of news on transnational Arab satellite networks. Injustices against Arabs are dealt with much more critically and extensively than injustices done to Israelis, whose victims are hardly present on screen. Al-Jazeera calls dead Palestinians 'martyrs' – a label reserved for Palestinians and not applied to dead Iraqis or other nationalities. The supremacy of a pan-Arab agenda evident in programmes broadcast by al-Jazeera becomes clear when the network – justifiably – criticises, time and again, injustices arising from Israeli or American policy and their militaries, while often downplaying the responsibility of Arab states and regimes and the role of privatised forms of violence (terrorism). Extensive reporting on the playing of video messages by Osama Bin Ladin have echoed terrorist messages.

In a nutshell, Arab media sometimes show the same tendencies as

Western media to subscribe to hegemonic nationalism. Some authors have legitimised that trend as a form of so-called 'contextual objectivity'.[10] The idea is that as long as American and sometimes European media are one-sided, Arab media have to be one-sided as well to balance out the Western dominance of international news flows and be heard in 'world opinion'. This approach is problematic.

Missing interdependence, or the deep structure of international reporting

Comparing case studies of framing and narratives in Western and Middle Eastern media reveals that intertextuality and dialogue in the foreign reporting on both sides exists but remains vulnerable and prone to nationalist predilections. 'Globalism' as a culture in political reporting is still underdeveloped, for a number of reasons. Firstly, mass communication cannot be dissociated from national interests. The theoretician of international relations, Richard Rosecrance, once argued that communication and conflict are linked in multiple ways.[11] If the underlying conflict between two countries is a zero sum game, as in the Palestinian-Israeli conflict where territorial gains for one side are losses for the other, communication is of minor importance since the problem is not about a lack of communication or a misunderstanding: the conflict is real and it will echo in different ways in the different media systems. It is only when politicians allow for a definition of the conflict as a 'win-win-situation' that communication can become more important. Which means: the so-called 'media dialogue' between the West and the Middle East can surely be no substitute for better policies in international relations. I think the real impact of the media on political decisions is sometimes vastly overstated.

Secondly, the fact that media echo real conflicts is itself the result of the fact that media systems are open to interference from a number of national forces and interests that limit their autonomy, since their major characteristic is still that of national media systems: structural deficits of Western media coverage of the Middle East and the Islamic world are mostly based on the fact that even in today's seemingly globalised world, foreign reporting is, by and large, determined by national (and sometimes regional) interaction between the media system and other sub-systems and social environments.

This pre-eminence of national over international interaction manifests itself in various ways. For example, there is a hegemony of national language communities creating their own long-term narratives of the world and overcoming them poses a real challenge to international understanding; national governments have refined their public relations instruments; since, in most cases, foreign reporting is predominantly directed at domestic audiences, national markets prevail over international markets and therefore foreign reporting often reproduces national moods, and if people, as they often do, 'rally round their flags', their media do too; and, probably as a result of the insulation of markets, the financial resources of the media are often very scarce in foreign reporting and the media become vulnerable to government PR.

National media systems might be interconnected in the sense that foreign correspondents and news agencies, in particular, are providing each nation with the raw material of information. But national media systems are not interdependent, since the way events are covered is not judged by those about whom the media systems report on – in this ease the people living in the Middle East and in the Islamic world – but by domestic audiences, who, due to their own distance from the matters reported, have hardly any means by which to judge the quality of the foreign news to which they are exposed. However, neither Western nor Middle Eastern media are so waterproof as to prevent many truthful facts on Middle East developments from entering the news. Also, at certain times the dynamics of public controversy allow for the so-called 'little traditions' of Oriental studies and expertise to find their way into the mainstream mass media in order to clarify public misperceptions. And as we have seen, the relative distance from involvement of a country in a heated international crisis or even in war can liberalise public debate.

Remember the case of British reporting during the Iraq War. That case is a bit of a conundrum to me, but one could start to think along the lines that perhaps the relatively pluralist British media coverage was a sign of slow and gradual Europeanisation – meaning that due to the process of integration into the European Union, the national media system might have lost its pre-eminence and opened up to more transborder influences from other European countries – and has done so despite all Euroscepticism in the country. However, this is mere speculation and we have no empirical evidence to support it.

However, that example should not push aside the fact that the West also contains 'endangered democracies'. Especially with respect to the biases in US mainstream media in times when the USA is at war, it is an illusion to contrast 'open' Western with 'authoritarian' Muslim or Middle Eastern media systems, since both sides can be utterly one-sided and the notion of plural independent democratic media is just an illusion. From a functionalist perspective on media systems that rids itself of the normative camouflage on media theory, it must be said that Western democratic media 'at war' can be as executive-friendly as any authoritarian media. The assumptions of contemporary philosophers about hegemony in culture and the authoritarian challenges of democratic development have long been proven correct by empirical social sciences.[12]

There is a need to come up with completely new suggestions for a new, more carefully designed and fair news management during international conflicts, starting with a reflection on the definition of responsible information management for governments which, like George Bush and Tony Blair, have often misinformed their people. The media must be provided with more resources for foreign reporting, and thought should be given to the setting up of real transnational media that are trusted by both Western and Middle Eastern audiences.

But what about this idea of 'contextual objectivity' that was mentioned before? What if the national and regional media are left to echo nationalist views of the respective peoples, in the hope that, due to global connectivity, the net sum of national media cultures creates something of a more balanced 'global public sphere'? I argue that such an approach is completely mistaken. The problem is that there is no such thing as a world media system since political and market interdependence is usually underdeveloped. This means, however, that 'contextual objectivity' cannot be a substitute for objectivity within national media systems which is still, as I see it, the main precondition for a media dialogue in European- or Western-Middle Eastern relations. Otherwise citizens in Europe, and more so in the USA and in the Arab world, are left alone with conflict interpretations that have a potential to fuel rather than moderate international conflicts.

There have been many fundamental misunderstandings about the global media in the course of the globalisation debate of the last ten to fifteen years. Many experts thought that direct satellite broadcasting would break up national media systems. But are we connected? Certainly not through direct access to the other side's media. Even where technical access is granted via satellite dishes, Western consumers hardly ever tune into the TV programmes of other countries.[13] Such access is limited to migrating minorities and very tiny information elites. We subscribe to a myth of globalisation if we assume that the world is construed like a giant web of communication that connects us all and comprises a 'global civil society'. On the contrary, satellite technology and the digital age have enabled the deepening and extension of geo-linguistic regional media spaces and intra-cultural media spheres – in Europe as much as in the Arab world. Arabs can tune into more than 150 TV channels. Europe itself is not as one because it comprises various language communities between which direct transborder effects, being almost non-existent, are hardly measurable.

The situation is a bit better with respect to Western media cooperating with Arab media – I say Arab media, although 99 per cent of Western journalists have certainly never heard of Pakistani or even Turkish TV channels, and they most probably know only one of more than a hundred Arab networks, al-Jazeera. Many Western media have cooperation agreements with Arab networks, therefore national Western journalism should be in a position to 'translate' Arab media discourses into Western programmes – but in fact it does not. I think there is an enormous misunderstanding when it comes to the impact of networks like al-Jazeera on the West. Western newsrooms usually do not hire Arabic-speaking personnel, and with few exceptions there is no continuous news watch. Globalisation has enabled Western media to select images from Arab TV, mostly of radicals like Bin Ladin, but on a daily basis the Western media have no real access to the media content of the Middle East. Perhaps the English version of al-Jazeera can make for a slight change, but I doubt it, because the difference runs deeper than pure language. Even the images that are being taken over by the West are highly selected, because there is still a deep divide between the imagery of Arab and Western, even European, media. You might not see many Israeli victims on Arab TV screens, but likewise Palestinian victims of Israeli violence in Western media are hardly ever as shocking as they are on Arab TV. There is still a deep gulf between the emotions conveyed by Western and Arab media. And as Alfred Grosser, the German-French-Jewish intellectual once said, peace can only be made when the suffering of 'the other' is understood and accepted. To that understanding, institutional collaboration between Western and Middle Eastern networks and media has so far contributed very little.[14]

It is my experience that the lack of awareness and the failure of what media scholars have branded as 'transnational media diplomacy' have many reasons; reasons that lie with the journalists, with news routines, the phenomenon of closed discourse, with the national priming of media systems and also with the orientation of owners and media institutions. One of the major impediments to connectivity and to intertextuality is the lack of transnationalisation of media capital. Media capital in the Middle East is mainly shared by Egyptian, Lebanese, Saudi Arabian and Gulf investors. Western capital in the Middle East is marginal and mostly confined to certain investments made by companies like Disney or Viacom.[15] Even a major transnational player like Rupert Murdoch does not generate more than 10–12 per cent of his capital from the Asian market and retains a clear focus on the non-Muslim parts of the Far East. The same is true for Arab capital in the West. As Naomi Sakr rightly concludes, transnational capital interdependence has not reached any socio-political depth and remains without political impact, since it is not only limited, but also 'blind' when it comes to imbalances in international reporting. The preliminary conclusion is that media diplomacy will receive little or no attention as long as media capital is allocated in either the West or the 'Islamic' world but not in both worlds at the same time.

Of course, in today's world there are numerous media that could have an impact on Euro-Middle Eastern relations. Unfortunately, the European Union, despite many debates within relevant circles, has never managed to set up a real transnational media culture within its own confines. Arte and Euro News or Euro Sport are about the only examples for this. But projects like the Euro-Turkish or Euro-Arab Arte have never reached completion. Remember the time when CNN, despite all its political and ideological shortcomings, represented the idea of a transnational TV for the world. Leaders from around the world spoke on CNN, they made politics there, and, as in the case of CNN's critique of America's support of Russian repression of Chechnya, their prominent position gave them the strength to mediate for peace. Today, CNN is but a shadow of itself, widely discredited in the Middle East for its pro-US reporting. If Europe wants to build a bridge between the Middle East and the West in the media age, it should probably remember that story of CNN.

There is another field in which the European potential for media dialogue is significant, although hardly anybody within Europe takes notice of it in foreign broadcasting. The BBC World Service, Radio France Internationale, Deutsche Welle, Radio Nederland and many others broadcast programmes in Arabic, Farsi, Turkish, Dari and Pashtu into the Middle East. Their advantage is that other than regular satellite broadcasting, those programmes actually often reach the monolingual middle and lower classes in the Middle East, and they are often very prominent. In Afghanistan, the BBC was the last station to host the Taliban leader Mullah Umar when the war broke out. In Iran, Deutsche Welle is a highly respected radio network, hosting the government as well as opposition forces. Many European foreign broadcasters follow a form that is clearly in opposition to American, Russian or Chinese hard-handed propaganda approaches. European programmes are often based on the idea of dialogue and try to compensate for the authoritarian shortcomings of Middle Eastern media systems. The problem with Europe is that its audiences and publics are not aware of this asset and that there are many bureaucratic hurdles to overcome. For example, Deutsche Welle is not allowed to air inside Germany, because that is considered the domain of the other public service broadcasters (although it is technically possible). This really gives an impression of how seriously handicapped the idea of Euro-Middle Eastern media dialogue still is. In theory, those networks would be an ideal platform for the exchange of political viewpoints and the intellectual integration of civil societies; in practice, they are still a one-way street.

Speaking of civil societies, what role does the Internet play for Euro-Middle Eastern relations? The average connectivity rate in the Arab world, except for rentier economies like the Gulf States, is around 3 per cent, albeit growing. Nevertheless it remains at a comparatively low level when compared with the West. Transnationality in the use of the Internet is really too complex to be grasped. There is certainly a long-term cultural international impact, yet the Internet is not primarily an international medium. Transborder connectivity is rising, but intra-border connectivity is growing even faster. Audience studies show that political information is among the priorities of heavy Internet users among the youth (more than ten hours per week), while it tends to be of much lesser importance among average and low level users.[16] Of course, that can turn out to be highly political in the end. But it has nothing to do with increasing knowledge of the Western international policies in the Middle East and vice versa. A study by the RAND corporation[17] argues that the Middle Eastern opposition groups are hardly present on the Net. As a result of severely limited connectivity in most Muslim countries, fax machines are still more important for political campaigning than the Internet. Those groups that are represented on the Internet, the study says, are mostly the opposition in exile and get more attention from Western audiences and from the UN system than from civil society in the Islamic world. Other experts are more optimistic that the Internet has enabled Middle Eastern NGOs to network with their counterparts overseas in a way that seems to have raised international awareness and strengthened pressure by Western NGOs on Arab governments.[18] However, the methodology of that literature raises doubts as to whether that judgement is sufficiently qualified, since evidence is mostly based on mere observation or, at best, content analysis of the homepages of Islamist organisations. Neither inter-organisational links nor the usage of these sites in either the Muslim world or the West have been researched in depth. The existence of net-based civil society networks is often presupposed, but hardly ever documented. The idea of opposition forces integrating and supporting each other across borders, called by some the 'Zapatista effect' due to the world-wide attention and support Mexican Zapatista rebels received, is good. Why should one subscribe to a narrow definition of international relations as inter-governmental? But, as in the mass media field, there seems to be a discrepancy between theory and practice that needs more research and attention.

If we were to try to summarise we could say, yes, intertextuality and dialogue have a chance on both sides, they have a certain history, but the connectivity level is still rather low and interdependence of media markets, organisations and policies is still underdeveloped, all of which cements the primacy of national or regional over global media systems and makes conflict reporting, in particular, vulnerable to 'warlords' of any kind, be they Middle Eastern terrorists or democratically elected governments. Huntington was wrong because there is no such thing as an unavoidable 'clash of Islam and the West'. But the media sometimes prove him right because in certain situations of international conflict they clearly give way to hegemony.

The media's impact on Euro-Middle Eastern relations

Bear in mind that we might be putting the cart before the horse if we ask the media to make possible a dialogue in international relations. Isn't it the task of politicians to do diplomacy, and aren't the mass media by nature unidirectional and non-interactive – a fact the German poet Bertolt Brecht already bemoaned in the 1920s after radio was introduced? But that was long ago and the media have changed and new public spheres have been opened up due to technical and social changes. So what is the real role of the mass media in international relations and in Euro-Middle Eastern relations?

Generally speaking, the mass media might be important for international relations in a broader, cultural sense of the term – for world views and patterns of activity of the individual European, Westerner, Middle Easterner, Arab or Muslim and for political opinion-making among citizens. But there is evidence for the fact that they are less important for international relations in the narrow sense of political relations and political elites. In any case, reflection is important because ordinary citizens' views might not be as significant for foreign as for domestic policy – think of immigration. And although the media might not be decisive for foreign policy much of the time, under certain circumstances they could be influential and there the media-policy game has certain rules.

Considering contemporary theory on the media-policy and on the media-citizen link with regards to foreign affairs, most authors would probably agree that the mass media are in a worse position than politicians, because politicians are the actors within international relations. Civil society is usually weakly organised when it comes to expressing views about world affairs, and the media are almost left alone with politicians who receive privileged information through a variety of channels from embassies to secret services and who can influence the substance and the dynamics of foreign policies and thereby create a near dependency. However, there are times when the media can gain superiority.

In Euro- and Western-Middle Eastern relations, the supremacy of politics over the media has been amply documented. In the Gulf War in 1991, the USA managed to prevent images of probably more than 100,000 Iraqi victims being shown on Western TV screens. They were able to control the media by creating information hubs, exerting censorship and public relations initiatives that were often outright propagandist and false – quite like in the 2003 war. Although many might say that, during the war in Iraq, the world media were critical of the USA, mainstream US media were not and that worked for George Bush. The supremacy of politics over media is not manifested in the fact that governments like that of Bush are not criticised, but in that they are able to regain the political momentum through dosages of either political-military activity or through political PR that 'trickles through' the media system because journalism is in most cases unable to check and dismantle disinformation as fast as would be necessary. Consensus is not needed – doubt is enough. As long as people doubt whether there could be danger for them or for the country, they let politicians go to wars, and as we have seen during active wartimes, even critical remarks are often taboo in the media. In several books and articles that were published in the last few years, the so-called 'CNN effect', the potential influence of media on foreign policies, has therefore undergone a critical reappraisal. Most authors say: The CNN effect tends to be a myth.[19]

Moreover, there is, as I see it, no evidence that the negative image Western media create of Islam has substantially influenced European or Western political relations towards Islamic governments or even towards Islamist movements. Just think of the good relations between the USA and countries like France with Saudi Arabia and Morocco, or even the very stable relations between Europe and Iran that have survived all problems. The USA has excellent relations with Islamist groups and parties in Algeria, Egypt, Turkey, etc. While the Western media image of Islam – not of the Middle East as a whole – is as bad as can be, Western foreign policies cannot be said to be driven by a culturalist complex of Islamophobia. It's realpolitik.

Of course, there are situations in which the media can have a greater impact. The literature has mentioned a few cases which are 'side stages' of

Western-Middle Eastern relations. Media are said to have been influential in maintaining the no-fly zones in Kurdish Iraq after the war in 1991 or in the formal break-up of the so-called 'critical dialogue' of the European Union with Iran in the mid-1990s. In these cases, it was the media that exerted pressure and changed policies while politicians had to follow. For the media to be effective in international policies, one or more of the following conditions must be fulfilled

There must be some kind of vacuum of decision-making, because different factions of politicians are moving in different directions. Since these splits happen to be bridged in wartimes, the media's influence is potentially stronger in times of a low-intensity conflict and left-wing political cleavages within the political elite are more outspoken. Alternatively, for the media to be influential, they can also base their coverage on a consensus within society, as is often the case with ongoing genocides, where most people would sympathise with intervention. Now, the last case is problematic, because, as the Darfur crisis – in which the United Nations decided not to intervene militarily – has shown, even 'genocide' seems open to definition and manipulation. For the media to exert pressure, they need one thing more than anything else: images of the humanitarian catastrophe. It is such images that can mobilise critical domestic audiences and exert pressure on politicians. Authoritarian regimes in the Middle East or even the US government in Guantanamo know that, and they try everything to prevent images of human tragedies leaving the country.

These rules for media influence also hold true for Middle Eastern media, despite the fact that European media are democratic and many Middle Eastern are at best half democratic. But, as I mentioned before, in the field of foreign reporting, these differences are often not as clear-cut. In the Middle East the human tragedies behind the wars in Afghanistan and Iraq, much more than any sympathies with the regimes of the Taliban or Saddam Hussein, led to widespread opposition towards US policy and, in the case of Afghanistan, also European policy. The human catastrophe of tens of thousands of uncounted victims was directly connected to the mood of resistance in the Muslim world. What was the role of the media in this situation? Was there something like an 'al-Jazeera effect' in the Middle East on politics and on public opinion? With regards to politics I doubt it very much. A country like Qatar could very well live with double standards and could allow its own TV station, al-Jazeera, to criticise the Americans while maintaining the largest American military base. The more the media aired the feeling that Afghanistan – and, more so, Iraq – were existential crises for the whole Arab world, the less Arab governments had to be afraid of being toppled by some kind of coup because also in the Middle East, people and the media tend to rally round the flag.

Now, if the media, in my very narrow definition of international relations in conflict, are rather weak, what about the media's impact on public opinion and on political mobilisation? With respect to media-citizen relations, modern theory has abandoned the old 'mood theory' of Gabriel Almond and replaced it by a multitude of mixed paradigms.[20] People have core values of war or peace or of solidarity with the Palestinians, let us say, that are not easily influenced by any media output. But, since mass media are covering distant events beyond the personal experience of most people, they often have an impact on people's political agenda, opinions and ideologies. For example, most Arabs might feel solidarity with the Palestinians and media telling them not to have such feelings would hardly be accepted and trusted. But, Arabs' choice of whether to opt for moderation or radicalisation in a certain political situation, whether they support political plans or a certain politician or party, can very well depend on the agenda and framing of the media. Both values and opinions create political action; values alone are not sufficient. And since foreign reporting operates at the crossroads of values and opinion, it is influential with regards to political opinions among the population. Opinions are formed by values manifesting within a certain media agenda. Audiences are powerful, but so is agenda setting and media framing. The so-called 'street' is as important as the media. Politicians know that – and that is the reason why they want to influence the media.

Media scholar Marc Lynch argued that the Arab media were operational in organising solidarity campaigns for Afghanis and Iraqis throughout the Arab world.[21] He argues that with the media, especially with satellite TV, a new public sphere has emerged. In that sense, Lynch holds that the media do play a role and that the 'Arab street' is becoming reactive to the mass media. I would support Lynch, but would also like to give a slightly different explanation. It is Arab media that can give Arab values a certain operative direction. Rather than claiming the superiority of the modern public media sphere over the 'street', one should be aware of the mutual relationship between people's values and the media agenda. In the case of Iraq, the Arabs' values were realised within the confines of the media agenda, which was itself limited by the rallying-round-the-flag syndrome wherein the media asked for solidarity rather than for a change in government.

When it comes to international affairs, the media in the Middle East have limited effects on national foreign policies, but they can be effective in shaping public opinion. The new sphere exists, it can reinforce public sentiments and shape political opinion, but it is hardly effective against those public sentiments. For almost three decades, Iranian mass media, including many of the most trusted papers of the reform movement of ex-President Khatami, have condemned the USA in almost everything they have done. Despite that, recent polls showed that 60–70 per cent of the Iranians would opt for better relations with the United States (the Iranian pollster was imprisoned for publishing these results). The results show that the media might have been successful in reinforcing a negative political image of America in a field close to people's values (Muslim solidarity, Jerusalem, etc.), but they are unable to operate against the seemingly cosmopolitan cultural interest of most Iranians and make Khomeini's critique of the USA and Western culture a generally acknowledged attitude. This reminds me of the East-West conflict and the inability of communist media to destroy the people's concept of 'the free West'.

Now, what about Europe and the West? I am not sure whether such positive pro-Western attitudes exist in the West. Polls show that 50–80 per cent of Western populations are frightened of Islam – and this sentiment is not confined to political topics. Probably as a result of the colonial history and the societal crises in the Muslim world, the contemporary Middle East seems much less attractive to the West than vice versa. While most Middle Eastern countries have always passed through times of greater cultural introspection and, at other times, open-mindedness towards the West – barely ever towards the East or to China, by the way – the West nowadays lives in a not-so-very-benign neglect of the 'East'. That is definitely not a culturalist statement of mine, saying that 'the West' ignores 'the East' in the way Edward Said did in his famous 1978 book, *Orientalism*. I mentioned at the beginning of this article that there was a time in the 1950s and early 1960s when the image of the Middle East in the West was more positive. But the history of Western news-making has allowed the media image of the Middle East to deteriorate enormously. It seems that media agenda has overshadowed Western public opinion of the Middle East.

If it is true that Western mainstream public opinion has currently little cultural resistance against Islamophobia, then the media in the West play a crucial role in activating or de-escalating such moods. When witnessing the organised outrage on the streets of Tehran against Israel and America, it was said that such public zealotry was completely unknown to the modern West. Although possibly true, such a view compares apples and oranges. If you consider that demonstrations such as those in Iran are a reaction to the feeling of many Middle Easterners of being politically re-colonised and hijacked by the West, and if you compare such feelings with the impression of many Europeans that they are being 'occupied' by immigrants, the comparison makes sense. A scientific report by the European Union argued that there are many trends of covert and overt racism in European media which are made partly responsible for open racism in European societies.[22]

We seem to be starting to understand how the media function in international relations. They are often weak vis-à-vis their own foreign affairs politicians, barely connected transnationally, but are an integral part of shaping the world views of their audiences. In the end, how do we answer the complex question of the role of media in Western-Islamic, Euro-Middle Eastern relations? Are we to emphasise the shortcomings of coverage, the danger of its influence – especially conflict reporting – on people's minds and world views, or the partly positive aspects in 'framing' the other, or the vulnerability of truthful and balanced reporting? Do we relax, because the impact of the media on international decision-making and on our politicians seems limited, or shall we work for an improvement of international reporting?

I suggest we admit that, with all reservations in mind, the media might not decisively influence people's world views today, but might be influential in the future and can influence political cultures in the long run. Even politicians grow up as ordinary people, and an improvement in media dialogue can surely help future policies to develop a really global political vision that challenges the Western-Middle Eastern divide.

Notes

1. Hafez, *The Myth of Media Globalisation*.
2. Huntington, *The Clash of Civilisations and the Remaking of World Order*.
3. Norris, *The Restless Searchlight: Network News Framing of the Cold War World*.
4. Ibid.
5. Hafez, *Die politische Dimension der Auslandsberichterstattung: Vol. 2: Das Nahost- und Islambild der deutschen überregionalen Presse*.
6. Hafez, 1996, 2000a, 2000b, 2000c, 2002.
7. Hafez, *Die politische Dimension der Auslandsberichterstattung: Vol. 2: Das Nahos- und Islambild der deutschen überregionalen Presse*, pp. 146–180.
8. Arendt, 'Truth and Politics', 'Lying and Politics'.
9. Pintak, *Reflections in a Blood-Shot Lens, America, Islam and the War of Ideas*, p. 72.
10. El-Nawawy/Iskandar, *Al-Jazeera: The Story of the Network that is Rattling Governments and Redefining Modern Journalism*, p. 54.
11. Rosecrance, *International Relations: Peace or War?*, p. 136 ff.
12. See Bocock, *Hegemony*.
13. Hafez, *The Myth of Media Globalisation*.
14. Grosser, *Media Ethics in the Dialogue of Cultures: Journalistic Self-Regulation in Europe, the Arab World, and Muslim Asia*.
15. Sakr, *Satellite Realms: Transnational Television, Globalisation and the Middle East*, pp. 69, pp. 93–96.
16. Ibrahrine, *New Media and Neo-Islamism: New Media's Impact on the Political Culture in the Islamic World*, p. 310.
17. Burkhart/Older, *The Information Revolution in the Middle East*, p. 38.
18. Sakr, 'Freedom of Expression, Accountability and Development in the Arab Region', p. 42.
19. Robinson, *The CNN Effect: The Myth of News, Foreign Policy and Intervention*, Gilboa, 'The CNN Effect: the Search for a Communication Theory of International Relations'.
20. Almond, *The American People and Foreign Policy*.
21. Lynch, *Voices of the New Arab Public: Iraq, Al-Jazeera, and Middle East Politics Today*, p. 156.
22. Ter Wal, *Racism and Cultural Diversity in the Mass Media: An Overview of Research and Examples of Good Practice in the EU Member States*.

Bibliography

Almond, Gabriel, *The American People and Foreign Policy*, New York: Praeger, 1950.

Arendt, Hannah, 'Truth and Politics', *The New Yorker*, 25 February, 1967.

Arendt, Hannah, 'Lying and Politics' *The New York Review of Books*, 18 November, 1971.

Bocock, Robert, *Hegemony*, Chichester: Ellis and London/New York: Tavistock, 1986.

Burkhart, Grey, E., Older, Susan, *The Information Revolution in the Middle East*, St Monica et al.: RAND Corp, 2003.

El-Nawawy, Mohammed and Iskandar, Adel, *Al-Jazeera: The Story of the Network that is Rattling Governments and Redefining Modern Journalism*, Cambridge, MA: Westview, 2003.

Gilboa, Eytan, 'The CNN Effect: the Search for a Communication Theory of International Relations', *Political Communication*, no. 22:1, 2005, pp. 27–44.

Grosser, Alfred, 'Journalism Ethics: The Mass Media's Role in Transnational Relations', in Hafez, Kai (ed.), *Media Ethics in the Dialogue of Cultures: Journalistic Self-Regulation in Europe, the Arab World, and Muslim Asia*, Hamburg: Deutsches Orient-Institut, 2003, pp. 33–35.

Hafez, Kai, 'The Algerian Crisis as Portrayed in the German Press: Media Coverage of Political Islam Communications', *The European Journal of Communication Research*, no. 21:2, 1996, pp. 155–182.

Hafez, Kai, 'Imbalances of Middle East Coverage: A Quantitative Analysis of the German Press', in Hafez, Kai (ed.), *Islam and the West in the Mass Media: Fragmented Images in a Globalising World*, Cresskill, NJ: Hampton Press, 2000a, pp. 181–197.

Hafez, Kai, 'The Middle East and Islam in Western Media: Towards a Comprehensive Theory of Foreign Reporting', in Hafez, Kai (ed.), *Islam and the West in the Mass Media: Fragmented Images in a Globalising World*, Cresskill, NJ: Hampton Press, 2000b, pp. 27–66.

Hafez, Kai, 'Transcultural Communication and the Antinomy between Freedom and Religion: A Comparison of Media Responses to the Rushdie Affair in Germany and the Middle East', in Thierstein, Joel and Yahya R. Kamalipour (ed.), *Religion, Law, and Freedom: A Global Perspective, Foreword by Cees J. Hamelink*, Westport, CN: Praeger, 2000c, pp. 177–192.

Hafez, Kai, *Die politische Dimension der Auslandsberichterstattung: Vol. 2: Das Nahost- und Islambild der deutschen überregionalen Presse*, Baden-Baden: Nomos, 2002.

Hafez, Kai, *The Myth of Media Globalisation*, Cambridge: Polity, 2007.

Huntington, Samuel P., *The Clash of Civilisations and the Remaking of World Order*, New York: Simon & Schuster, 1996.

Ibahrine, Mohammed, *New Media and Neo-Islamism: New Media's Impact on the Political Culture in the Islamic World*, Saarbrücken: VDM Müller, 2007.

Lynch, Marc, *Voices of the New Arab Public: Iraq, Al-Jazeera, and Middle East Politics Today*, New York: Columbia University Press, 2006.

Norris, Pippa, *The Restless Searchlight: Network News Framing of the Cold War World*, Cambridge: Cambridge University Press, 1995.

Pintak, Lawrence, *Reflections in a Blood-Shot Lens, America, Islam and the War of Ideas*, London/Ann Arbor, MI: Pluto, 2003.

Robinson, Piers, *The CNN Effect: The Myth of News, Foreign Policy and Intervention*, London/New York: Routledge, 2002.

Rosecrance, Richard, *International Relations: Peace or War?*, New York: McGraw-Hill, 1973.

Sakr, Naomi, *Satellite Realms: Transnational Television, Globalisation and the Middle East*, London/ New York: I. B. Tauris, 2001.

Sakr, Naomi, 'Freedom of Expression, Accountability and Development in the Arab Region', *Journal of Human Development*, no. 4:1, 2003, pp. 29–46.

Ter Wal, Jessika (ed.), *Racism and Cultural Diversity in the Mass Media: An Overview of Research and Examples of Good Practice in the EU Member States, 1995–2000*, Vienna: European Monitoring Centre on Racism and Xenophobia, 2002.

ENRIQUE KLAUS

The Integration of Weblogs in the Egyptian Media Environment

In the latter half of Hosni Mubarak's reign the Egyptian media environment witnessed profound changes induced by the appearance of new operators. By the end of the 1990s, the success of satellite TV stations and the emergence of a private press sector were the most salient elements of the observed transformations of the Egyptian news market. In addition to these new media, we should mention the weblogs that progressively imposed themselves in public debates to become an integral part of the national media operators.

This chapter discusses the integration of these new operators into the Egyptian media environment. The underlying idea is to provide a localised view – that of the Egyptian news professionals – on the social phenomenon of weblogs in Egypt. In other words, from an examination of the convergence of interest between weblogs and newspapers, I analyse here a particular form of 'globalisation' understood as 'the local production and consumption of information, which is to a large extent formatted and conveyed through the networks of the globalised system'.[1]

From blogosphere to bloggers: does a media community exist?

Since their first appearance in 2004, the number of Egyptian weblogs has been incessantly increasing: by the end of 2006, the daily *al-Ahram* estimated the number of Egyptian bloggers at 5000 while one of the pioneers of the blog in Egypt, 'Ala' – 'Abd al-Fattah – spoke of some 1200 bloggers[2]. The difference between these numbers shows the difficulty of making an inventory of blogs; this can be attributed to the variety of criteria used for these estimates. One can include, or not, the Egyptian bloggers residing abroad and one can count, or not, the weblogs relaying news items made of personal events and addressed to an audience restricted to a circle of friends. The difficulty rises with the existence of blogs that are not kept up to date and the appearance of others created in a circumstantial manner in relation with an event; like the Kifaya Heroes blog that was run by three bloggers from the opposition movement for two months in 2006 in order to relay information about their friends' detention, or the *Ana Harami* (I am a thief) blog, created at the time of the Annapolis Conference in November 2007.[3]

Early in the Egyptian blogosphere's history, attempts to federate bloggers were observed whether by means of blog aggregators (manalaa.net and later omraneya.net), or through the project of creating a sub-branch inside the journalists' union. While launching a second campaign to unite Egyptian bloggers, 'Ala' 'Abd al-Fattah and his wife Manal Hasan mentioned the project of the creation of a popular press based on a network of mostly Cairene bloggers, constituting 'the Egyptian blogosphere' (*halqat al-mudawwinin al-misriyyin*), according to their own words. The term 'blogosphere' is a trendy neologism, which designates, in Egypt as elsewhere, all or part of the individuals regularly expressing themselves through a personal weblog. Since it is a tool operating on a global network, attributing a nationality to it clearly participates in reframing the phenomenon nationally, i.e. consistent with political and social issues linked to freedom of expression as they pose themselves in the context of the Middle East in general and in Egypt in particular. The expression 'Egyptian blogosphere' implies the existence of a community linked and defined by the use of weblogs on a national scale. It thus transcends the religious (the Coptic or Bahai blogs), political (Muslim Brotherhood sympathisers, socialist revolutionary militants, or even the activists from the Youth for Change Movement and their respective blogs) and linguistic (Arabic, English or bilingual blogs), pertaining to that which is observable in the so-called blogosphere and, not without difficulties, also in the political and social life of contemporary Egypt.

Are the outcomes of this unifying policy of the Egyptian weblogs – the aggregator Omraneya coming online at the end of 2006 – and the transcending of community lines, as claimed by the Egyptian users of this global media tool, suffice to argue that the bloggers constitute a community? Do they not rather represent, beyond their discourse, a simple collection of individuals who carry out their blogging activities isolated from one another?

Certain elements of the short history of the weblog in Egypt are relevant for an analysis of the blogosphere as a community on and off the screen. As both a new media and a new mode of expression, the weblog requires its users (blogger or visitor) to master some of the codes and conventions of virtual communication. Mastering these modalities highlights specific competencies – beyond technical requirements – that the blogger must display to express himself online. Be it through posts or comments, bloggers and readers often handle dialectical Arabic, the main language used on the weblogs, in a quite unconventional way. Other than the required linguistic proficiency, dialectical Arabic transcription (in Arabic or Latin characters)[4] is also a particular skill developed by the members of the blogging community.

Regular weblog reading also sheds light on the reticular functioning of the online relationships between bloggers. This networked community is indexed through the hypertext links to be found in the margins of the webpage, which connect to the blogs that the blogger likes to read, to those of his friends or of simple acquaintances. Bloggers' comments on the pages of other bloggers also make these networks visible. In an even more obvious manner, the series of arrests of bloggers in 2006, mostly related to 'offline' activism, resulted in diverse solidarity movements transcending the most entrenched political rivalries. Thus, Karim Amir, the first blogger to be sentenced for his online writings and his opinions, which were judged to be blasphemous, received the support of bloggers originating from diverse leftist movements as well as some bloggers from the Muslim Brotherhood.[5]

Beyond the screen, one observes that online interactions between bloggers often operate in parallel with 'real' social relationships between them, especially when bloggers also share militant activities. The simultaneous emergence of weblogs in 2004 and the Kifaya opposition movement, on top of the enlistment of certain bloggers in its ranks, certainly enhanced a feeling of community among them. The demonstrations that the movement organised offered a 'real' space for these individuals to meet (the traditional Cairene demonstration places) and a specific context (that of police brutality) for shared experiences. Between 2005 and 2006, a time when protest marches occurred regularly, a dynamic emerged between the 'virtual' and the 'real' world of opposition. The demonstrations were first announced on the Kifaya site and then on a number of blogs that offered all the practical details of how to get to the location (public transport itinerary with Google earth maps). In the few hours that followed the dispersion of the demonstration, numerous blogs offered up-to-date pictures, anecdotes and first-hand accounts of the events of the demonstration.

Beyond the aggregation, the solidarity and sociability observable in the aforementioned examples, one would not fail to question this idea of community, especially when it comes to individuals who certainly interact but do so with no physical co-presence, through a medium. The fact that the titles of numerous weblogs mirror the individual character of the undertaking of creating a blog[6] invites us to handle carefully the seductive expression 'Egyptian blogosphere'. Indeed, a weblog is first and foremost a tool that individuals – and, less frequently, groups of individuals – use with various aims. The criteria for the choice of information relayed in these online spaces vary greatly from one blog to another. Some among them are dedicated to personal events shared in a recreational perspective with a personal social network. Others aim exclusively at subjects of more general interest, such as the political, economic and social situation in Egypt or in the Middle East. Still others, in the same line, explicitly refer to a vocabulary borrowed from the field of journalism ('chief editor', 'online journal', 'report'), organise their interface much like a newspaper's paste-up and include regular columns. In other words, the latter display a proto-professional competence to report facts of an interest largely surpassing the simple networks of personal acquaintances.

Al-Waʿi al-Misri (the Egyptian conscience), the blog that Waʾil ʿAbbas launched in February 2005, offers a good example of this last type of blog. Almost every week, some twenty-one contributors express themselves on *al-Waʿi al-Misri* and tackle current issues from Lebanese-Syrian relations to the lack of legal parity in matters of adultery, or even the retreat of the

Egyptian opposition parties from the 'national dialogue' advocated by the regime.[7] The blog was initially organised on a mailing-list model, close to the one that Wa'il 'Abbas formerly held. A month after the blog's creation, pictures of demonstrations (in commemoration of the beginning of the American occupation of Iraq)[8] took over the written content of the blog which was reduced to short editorials signed by the blogger, image captions, calls for demonstrations and news releases of diverse Kifaya sub-groups. The visual content took paramount place: caricatures and photomontages were introduced little by little and the blogger, then correspondent for a German press agency, proceeded to publish his own photo-reportage on different events such as the Sharm al-Shayk bombing, the violent repression of the Sudanese refugees' occupation movement in Cairo, and even the official ceremony for Naguib Mahfouz's funeral.[9] Wa'il 'Abbas does not use the word 'reportage' and prefers the term 'coverage' (*taghtiyya*) to signal the specificity of his take on the Egyptian media environment and, at the same time, his insertion into it.

With regard to our line of questioning concerning the integration of blogs into the Egyptian media environment, this type of weblog with pseudo-journalistic ambitions is of more specific interest to us. In the literature that is starting to be dedicated to blogs in the Arab world, there are diverse methods to classify them: some contrast the so-called political blogs to so called personal blogs, at the risk of forgetting that blogging is above all a personal activity and that some blogs largely dedicated to their authors' private lives can suddenly find themselves at the front of the media scene.[10] Others distinguish, in a more judicious and less rigid manner, activist blogs (cyberactivism) from bridgebloggers (blogs written in English addressing mainly an extraterritorial audience) and public sphere bloggers (Arabic language blogs that cover current events, while this political interest is not pursued in an activist commitment).[11]

At the cost of leaving aside the weblogs of specifically egocentric orientation, I for my part prefer to distinguish blogs on the field of journalistic work with a more or less professional style of reporting facts. Whatever the degree of professionalism they demonstrate, I am interested here in the individuals who adopt an approach, in their regular activity as blogger, that makes them occasional or regular operators in the Egyptian news market. From this standpoint, it is necessary to trace the brief history of the integration of bloggers into the arena of traditionally used media.

The intrusion of weblogs into the traditional media public sphere

It is impossible to separate the history of weblogs in Egypt from the recent political history of the country, and notably, that of the Kifaya movement. The first weblogs made their appearance in 2004, the year the movement started to organise, initially around the opposition to the hereditary transfer of power. A non-negligible number of bloggers joined the movement that then channelled diverse trends of the Egyptian opposition. Parts of the opposition press that followed the activities of Kifaya viewed the movement favourably. The bloggers of the movement, as such, made the news for the first time with an event that occurred during the protest demonstrations organised by Kifaya on the 25 May 2005, the day of the referendum to amend the constitution to allow presidential elections. On this day the security forces, assisted by thugs in civilian clothing, violently repressed the demonstration and took on a female journalist covering the gathering in front of the journalists' union headquarters. Men in civilian clothing harassed her, tore her clothing off and molested her. Activists photographed the scene and the images circulated on a number of weblogs before being published in the opposition press.

This episode that inaugurated the entry of bloggers into the Egyptian public sphere situated them in the camp of the opposition movement rather than depicting them as operators in the media environment. Similarly, on 7 May 2006, the arrest of at least five bloggers among other Kifaya activists on the sidelines of a demonstration in solidarity with the reform-minded judges rekindled the interest of the independent press in these individuals, still in their status as activists. These arrests were the first efforts of the regime to intimidate the bloggers who, beyond the arbitrariness of the arrests during the demonstration, seemed particularly targeted. Among them, the case of 'Ala 'Abd al-Fattah, RSF-Deutsche Welle award winner in October 2005, and that of Muhammad al-Sharqawy, beaten and subjected to torture of sexual nature while in custody, were particularly covered by the opposition press.

Fairly quickly, by the time of the elections in autumn 2005, some bloggers seemed to be competing with the journalists on their own professional field. During the electoral campaign for the presidential election, as highlighted by the regional daily *al-Hayat*, Wa'il 'Abbas distinguished himself through his blog being the only media, besides the al-Jazeera website, to present election polls.[12] On the day of the vote, a number of blogs published photographs of various irregularities that fractured the electoral process and ruined the image of the 'great march towards democracy' lauded by pro-regime newspapers.[13] This ongoing competition of bloggers on journalistic terrain passed almost unnoticed in the Egyptian newspapers: only *Akhbar al-Yawm* conceded that bloggers had invited themselves to the presidential election, but only as 'observers of the electoral process'.[14] In return, the journalistic potential of the weblogs was noted, after *al-Hayat*, by other extra-national media, testifying in this case to a certain conservatism of the national media when faced with the smooth entry of these new operators. The evening of the vote, 7 September 2005, al-Jazeera aired a special programme with Muhammad H. Haykal as star guest. In the immediate post election analysis, Haykal declared:

> There is a new phenomenon on the Internet, called 'blogging'. In the Internet society, the one that we don't see, I have found somebody who writes with a borrowed name, Bahiyya. I don't know who this Bahiyya is, but I swear that I ask my secretary to get me her articles. As soon as she writes an article I read it with more thought and respect than for any journalist I read in the press.[15]

This anecdote would appear almost obsolete if it had not been systematically recounted in newspaper articles[16] interested in the weblog as a social phenomenon. In the press exegesis of the blogosphere the anecdote takes the place of the founding act of the relationship between blogs and traditional media and seems to give the bloggers a professional free hand, as it involves Haykal, a great name in contemporary Egyptian journalism.[17]

In the wake of the presidential election, the relationships between journalists and bloggers were quite complicated and the gates of the 'legitimate' media environment, that of the news professionals, were not wide open to these newcomers. During the legislative election, several bloggers wrote and published numerous reports about bus convoys for the 'parachutist' voters to the risk districts (obviously, where a Brotherhood candidate was in a position to win) and about the brutality of the police and the NDP thugs preventing the non-supportive voters from reaching the ballot boxes. The quality of some pictures was visibly appreciated by print newspapers that cheerfully drew from this new source without bothering to credit the photographs published on their frontpages. The bloggers took offence, and in the middle of the elections, Wa'il 'Abbas went on strike to protest against the usurpation of his copyrights.[18] In order to cover the end of the elections he only published, at first, agency photographs and photographs from '[his] sources from the Muslim Brotherhood', before embedding a transparent site logo over the whole surface of the images he put online.

The elections of 2005 thus marked a stage in the slow integration of weblogs into the Egyptian public sphere. On this occasion, the originality of their contribution lay less in the revelations of violence and irregularities, which had been an integral part of earlier electoral processes, but rather, in their documentation. In fact the blogs' interface is so unique and original that it allows, on the same platform, the association of several modes of documentation, if not of verification, of information relayed through text, video or imagery. It is in this extent that they present an advantage over older media and, especially, over the written press with which they are henceforth in direct competition. It is also in this that, at the time of the 2005 elections, the bloggers distinguished themselves from the rest of the media sector.

With these first exploits, the bloggers cemented their presence in the Egyptian media environment by pursuing this innovative enterprise of documentation of the news, outdoing the other operators which are constrained by the periodicity of their media. By the end of October 2006 several bloggers again hit the headlines by revealing a moral scandal which took place in the centre of Cairo: five blogs recounted with considerable detail attacks against passers-by who supposedly were undressed and sexually abused by groups of men without the police intervening.[19] For five days not a single media source mentioned the incident; the delay and the detour to the second media 'revelation' shed light on the compartmentalisation of the different media operators. The dailies then took recourse to large extracts of weblogs and published some of their photographs.[20] The partisan weeklies were caught unaware and had to wait three to five days to give their headlines on the subject.[21]

From the first articles, the polemic dealt with the veracity of the facts and, beyond that, with the reliability of blogs in matters of information. This episode was the chance to revisit the editorial and political conflicts between different types of news media. After rejecting the story outright, 'considering that the source of the story is not a source of confidence', the pro-regime press did its best to accuse the bloggers of spreading rumours and of perpetuating the 'crime of falsification'.[22] The bloggers found support in the non-partisan independent press, which seized the occasion to accuse the regime, and particularly the Ministry of the Interior, of assuring only the security of those in high places in the government, at the expense of the security of the people.

Shortly thereafter, in November 2006, a second incident put the bloggers at the forefront of the media scene again. The video of a torture session in a Cairene police station circulated on several blogs before being relayed by Wa'il 'Abbas' blog. He specified in the caption, 'It is what we have heard of, from the mouths of some activists who have endured it. This is what happens in police stations and in political prisons. And now we see it with our own eyes.'[23] It was yet another video denouncing the degrading treatment administered in custody, that was published by bloggers. But this time, the precedent of the sexual harassment affair allowed the media to react efficiently: the 'independent' weekly *al-Fajr* launched a campaign to find the victim and, as often with this type of event,[24] to 'de-singularise' the incident in order to upgrade it to the level of a cause against 'a secret torture organisation from the Ministry of Interior'.[25] The campaign that was run by *al-Fajr*, with other national and international papers, bore fruit and led to the arrest and subsequent sentencing of the torturers.

Numerous are the bloggers who regularly denounce police brutality against demonstrators and the abuses that the police officers inflict upon them in detention, especially since the testimony of Muhammad Al-Sharqawy. Some blogs, the best known of which is *Tortureinegypt*, dedicate their editorial line to the denunciation of this type of violence. With the issue of the video, and the testimonies of the bloggers, it is really the documentation of this type of widely known misdeed that is new and, without doubt, parallels the sordid habit of some torturers to record their abuses and to circulate the videos as a warning.

Finally, to complete this short history, we need to mention the particular interest of the press for a sub-group among the bloggers, composed of those displaying an affiliation with the Muslim Brotherhood. The breakthrough of this second generation of bloggers, between the end of 2006 and the beginning of 2007, could be seen as directly linked to the wave of arrests that followed the paramilitary parade of the student Brothers at Al-Azhar University. Some of the fifty bloggers who had placed the Brotherhood's emblem on their homepage situated themselves in a double process, simultaneously an activist process, by means of their support for the leaders of the Brotherhood who were incarcerated at the end of 2006, and a testimonial one, aiming to convey another image of the Brotherhood.[26] In general these bloggers do not display their influences in the proto-journalistic practices of blogs, with the notable exception of 'Abd al-Min'im Mahmud, blogger and journalist for several traditional media organs.[27] But the foothold they constitute for the hardly loquacious Brotherhood and their opposition to the status of a party structure for the Brotherhood have not escaped the Egyptian political commentators and, particularly, the national press.[28]

Provisional balance of the interactions between bloggers and traditional media

The few episodes described here show the increasingly regular interaction between media from diverse generations and the growing integration of the weblog in the media environment. We could observe that the insertion, if sporadic and limited, of weblogs in the public sphere espouses the political and editorial divisions that existed before their arrival in the worlds of the Egyptian press. This new media situation has its consequences for the technical and editorial conception of different types of media of diverse forms to integrate weblogs, and, therefore, to structure the public sphere. Here I want to risk a (very) provisory appraisal of the changes observed in the last three years.

Professionals in the news market adopt three types of attitude towards blogs: media conservatism, integration into the market and media adaptation. The first type of attitude, characteristic for the publications of the pro-regime press, consists of refusing all integration of new media. This refusal does not only concern weblogs. The regular diatribes of *Ruz al-Yusif* regarding satellite TV stations in general and al-Jazeera in particular – which is accused of being an accomplice of the Muslim Brotherhood – are there to remind us that media conservatism concerns all extranational media. In the case of the weblogs more specifically, this conservatism can take different forms and, above all, brings out the non-professional and activist character of its authors. *Al-Ahram*, for example, practises a form of censorship of all information relayed by weblogs, notably so at the time of the video of the torturing in the police station. In addition, this institution distinguished itself a couple of months earlier (in August 2006) by blocking access to weblogs from the newspaper's own network before revoking the ban a week later as a result of a collective protest from the paper's journalists. *Ruz al-Yusif* presents itself as the editorial spearhead of the struggle against transnational media: as early as 2003 the paper had denounced the interference of the Internet as dangerous for Egyptian sovereignty in its own news market, and warned against the 1211 mailing lists of different Egyptian religious and political groups which were listed on Yahoo at the time.[29] Three years later its chief editor, Karam Gabr, launched denigrating campaigns against the bloggers at the slightest opportunity, notably during the sexual harassment incident.

The other two attitudes, integration in the market and media adaptation, are more common in the generally non-partisan opposition press. The first aims at inserting the weblogs into the news economy labelled as sources and/or political activists. This is most common among the so-called independent dailies because of the recently increasing competitiveness in this sector[30] and because of the structural constraints of editorial work in a daily publication.

Finally, media adaptation can take different forms depending on which type of media is setting about it. On parts of the satellite networks, faced with the success of online videos on YouTube, al-Arabiya and then al-Jazeera launched their own sites hosting amateur videos. In June 2006, al-Jazeera equally created a blog, al-Jazeera Talk, to which correspondents collectively contribute. Individuals writing their own blogs are offered the chance to publish reports and information on the network's blog after a brief training in professional ethics. In the case of the written press, *al-Dustur* is the publication that went the farthest in its adaption by creating, in August 2007, a weekly column dedicated to Egyptian and Arabic weblogs and a blogger, 'Abd al-Min'im Mahmud, was recruited as journalist for the daily paper. More anecdotally, we could mention the observation that the internet sites of a number of opposition papers received more attention, with notably more attractive interfaces, the ability to download articles in PDF format from the archives, or the embedding of a news ticker.

On the part of the bloggers, interaction with the world of professional journalism has led to a new interest for thematic press archiving, sometimes in an activist manner, following the example of the site id3m (*id'am*, support). This recent integration had the most repercussions for the bloggers with a proto-professional approach. These more than others base the success of their blogs on regular updating and must, like journalists,[31] 'find topics' or think about 'terminating ones weblog because of the psychological state that accompanies the feeling of not having anything new to present.'[32]

The various interactions with traditional media have also entailed some structuring in the blogosphere. Because of the profoundly individual character of the weblog, the transfer of information from blogs to other media is often based on the personal relationships between journalist and active bloggers. We could also observe hierarchies among bloggers, notably in the torture incident, and the emergence of 'bridgeheads' – generally proto-journalistic blogs – between lesser-known blogs and media.

Observing the trends indicated by the statistics measuring hits on weblogs, one can see the impact of interaction with the traditional media. The highest numbers of visits to most of the blogs cited here correspond clearly with incident-related interest of traditional media in information relayed by the blogs. The public sphere of the traditional media functions like a resonating field for the bloggers, who seem to need a double coverage of the information they relay to have an impact of any kind in the Egyptian public life and to be fully operational in an increasingly competitive media environment. The diverse forms of blogs' integration into the media presented here move towards recognition and a growing importance of bloggers in the Egyptian public sphere.

Overall, the configuration of the Egyptian public sphere is being altered by the integration of this new type of media. On one hand, the activist dynamism of certain bloggers and on the other, the conservatism of parts of the journalistic domain, give the impression that there are two parallel arenas in the Egyptian public sphere. The publications accredited by the

High Council for the Press and other traditional media are situated in the 'legitimate' arena of the public space, that of the news professionals. In this regard there is also an unofficial arena, not recognised as such, constituted of individuals and nourished by individual experiences. As such this arena is closer to the oppositional public sphere of Oskar Negt[33] than to the rationalist version of Jürgen Habermas.

Notes

1. Yves Gonzalez-Quijano, '*Sammi' sawtak!* les sites d'information en ligne dans l'espace public syrien', p. 70.
2. *Al-Ahram*, 6 December 2006 and cf. Intervention of the blogger, Al-Sawy Web Salon, 7 December 2006. Marc Lynch, 'Blogging the New Arab Public', speaks of a little more than a 1,000.
3. http://kefayaheros.blogspot.com/; http://anaharame.blogspot.com/.
4. The trancription of Arabic happens with the help of a syncretic alphabet developed in the use of SMS and consists of Latin letters and Arabic numbers to transliterate the letters with a pronunciation that is unique to Arabic. According to the conventions most used in Egypt, '2' tranlitterates the Hamza and the ellipse of the Qaf in the Egyptian dialect. '7' replaces ha' and '3' 'ayn. The emphatic letters are not signalled.
5. As a result of his posts criticising the religous education in Al-Azhar, Karim 'Amir, also known as 'Abd al-Karim Nabil Sulayman, was arrested on two occasions in 2006 and sentenced to four years in prison for contempt of Islam and insulting the president of the Republic. A support committee was formed. http://www.freekareem.org/
6. Apart from those using their name (Sharqawy, Manal and 'Ala'), numerous bloggers chose one or more elements of their identity which testify to the clearly introspective and individual aim of the weblog: 'I am like this' (Ana Keda), ' Thus I am' (Hakaza Ana), 'I am a (Muslim) Brother' (Ana Ikhwan), 'An Egyptian (woman)' (Wa7da Masrya) or even 'An Egyptian girl' (Bent Masraya).
7. *Al-Wa'i al-Misri*, respectively of 4 and 17 February 2005, 25 March 2005 and 24 February 2005.
8. *Al-Wa'i al-Misri*, 15 March 2005.
9. *Al-Wa'i al-Misri*, 29 July 2005; 31 December 2005, 31 August 2006.
10. This is notably the case of Radwa Usama (http://hakazaana.blogspot.com/) at the time of the incident of the 'sexual madness' (infra). For analysis of an Egyptian blog following an ego-centred editorial line and introducing the distinction between 'political blog' and 'personal blog' see George Weyman, 'Speaking the Unspeakable: Personal Blogs in Egypt'.

For an analysis of political Egyptian blogs see Rania Al Malky, 'Blogging for Reform: the Case of Egypt'.
11. Marc Lynch, 'Blogging the New Arab Public'
12. *Al-Hayat*, 29 August 2005. The results of the survey can be seen on *Al-Waʿi al-Misri* from 7 September 2005.
13. Notably *Manalaa.net*, *Tortureinegypt* and *Al-Waʿi al-Misri*, 7 September 2005.
14. *Akhbar al-Yawm*, 27 August 2005.
15. A transcription of the whole show is available at http://www.aljazeera.net/NR:exeres/6C37747E-4CCD-9AB9-BD1A9B7E5715.htm
16. *Al-Sharq al-Awsat*, 5 February 2006; *Al-Hayat*, 16 April 2006; *Al-Fajr*, 6 June 2006; *Al-Ahram*, 6 December 2006.
17. The professional reputaion of Muhammad Hasanayn Haykal is based on his many experiences. His career started in 1942, as a war correspondant. He then became a journalist at the parliament before taking on the managment of the national press under Nasser, with whom he collaborated closely (Muhammed Haykal *Bayn al-sahafa wa-l-syasa*). In 1971, under Sadat, he fell into disgrace and reappeared only at the end of the 1990s on al-Jazeera, where he hosted a show dedicated to the contemporary political history of the Middle East (*maʾa haykal*).
18. *Al-Waʿi al-Misri*, 18 November 2005.
19. http://ma3t.blogspot.com/, 24 October 2006; http://malek-x.net/ et http://t5at5a.blogspot.com/, 25 October 2006; http://hakazaana.blogspot.com/ et http://www.speaksfreely.com/, 26 October 2006. For more detail about this incident see Enrique Klaus, 'The ʿAyd 'Sexual Rage Scandal': Managing the Lack of Information through Categorical Photo-Fit'.
20. *Nahdat Misr*, *Al-Misri al-Yawm* and the weekly *Al-Karama*, 31 October 2006; *Al-Dustur*, 1 November 2006.
21. *Al-ʿArabi*, 5 November 2006; *Al-Fajr*, 6 November 2006; *Al-Usbuʿ*, 7 November 2006.
22. *Ruz al-Yusif*, 31 October 2006, also *Ruz al-Yusif*, 3 November 2006 and 12 November 2006; *Al-Ahram*, 1 November 2006.
23. *Al-Waʿi al-Misri*, 9 November 2006.
24. Cf. Luc Boltanski, *L'Amour et la Justice comme compétences, trois essais de sociologie de l'action* and Luc Boltanski and Elisabeth Claverie, 'Du monde social en tant que scène d'un procès'; Elisabeth Claverie, 'Procès, affaire, cause: Voltaire et l'innovation critique' and Elisabeth Claverie, 'Apparitions de la Vierge et 'retour' des disparus. La constitution d'une identité nationale à Medjugorje (Bosnie-Herzégovine)'.
25. Cf. *Al-Fajr*, 27 November 2006.
26. Marc Lynch, 'Young Brothers in Cyberspace'.
27. Correspondent for *al-Hiwar* and *al-Jazeera talk*. ʿAbd al-Minʿim Mahmud

directs three blogs (http://m-monem.blogspot.com/; http://monem-press.blogspot.com/; http://ana-ikhwan.blogspot.com/) and writes the column 'blogs' in the weekly *al-Dustur*.
28. Notably *Majalat Ruz al-Yusif*, 1 November 2007.
29. *Ruz al-Yusif*, 5 December 2003.
30. End of 2007 the market for dailies had five newspapers of the national press, two partisan papers and four so-called independent newspapers.
31. Sune Haugbolle, 'From A-List to Webtifada: Developments in the Lebanese Blogosphere 2005 – 2006', makes the same argument concerning Lebanese blogs.
32. http://molotoofy.blogspot.com/; 14 August 2007.
33. Oskar Negt *L'Espace public oppositionnel*.

Bibliography

Al-Malky, Rania, 'Blogging for Reform: the Case of Egypt', Arab Media & Society, February 2007. On line: http://www.arabmediasociety.com/topics/index.php?t_article=39/

Boltanski, Luc, *L'Amour et la Justice comme compétences, trois essais de sociologie de l'action*, Paris: Métailié, 1990.

Boltanski, Luc and Élizabeth Claverie, 'Du monde social en tant que scène d'un procès', in Luc Boltanski, Élisabeth Claverie and others, eds, *Affaires, scandales et grandes causes: de Socrate à Pinochet*, Paris: Stock, Les Essais, 2007.

Claverie, Élizabeth, 'Procès, affaire, cause: Voltaire et l'innovation critique', *Politix*, 1994, p. 26.

Claverie, Élizabeth (2002), 'Apparitions de la Vierge et 'retour' des disparus. La constitution d'une identité nationale à Medjugorje (Bosnie-Herzégovine)', *Terrain* 38, 2002. Online: http://terrain.revues.org/document1912.html/

Gonzalez-Quijano, Yves, '*Sammi' sawtak!* les sites d'information en ligne dans l'espace public syrien', in Yves Gonzalez-Quijano and Ch. Varin, eds, *La société de l'information au Proche-Orient: Internet au Liban et en Syrie*, Beirut: Université Saint-Joseph, 2007.

Haugbolle, Sune, 'From A-List to Webtifada: Developments in the Lebanese Blogosphere 2005 – 2006', *Arab Media & Society*, February 2007. Online: http://www.arabmediasociety.com/topics/index.php?t_article=91/

Haykal, Muhammad Hasanayn, *Bayn al-sahafa wa-l-syasa* [Between press and politics], Cairo: Dar al-Shuruq, 2003.

Klaus, Enrique, 'The *'Ayd* 'Sexual Rage Scandal': Managing the Lack of Information through Categorical Photo-Fit', in Enrique Klaus and C. Hassabo, *Egyptian Chronicles 2006*, Cairo: CEDEJ, 2007.

Lynch, Marc, 'Blogging the New Arab Public', *Arab Media & Society*, February 2007. On line: http://www.arabmediasociety.com/topics/index.php?t_article=32/

Lynch, Marc, 'Young Brothers in Cyberspace', *Middle East Report* (245), winter, 2007. On line: http://merip.org/mer/mer245/lynch.html/

Negt, Oskar, *L'Espace public oppositionnel*, Paris: Payot, 2007 (1971).

Weyman, George, 'Speaking the Unspeakable: Personal Blogs in Egypt', *Arab Media & Society*, October, 2007. Online: http://www.arabmediasociety.com/topics/index.php?t_article=164/

Indicative Webography
http://www.alexa.com/
http://www.aljazeera.net/
http://www.aljazeeratalk.net/portal/
http://anaharame.blogspot.com/
http://ana-ikhwan.blogspot.com/
http://43arb.info/meit/
http://baheya.blogspot.com/
http://www.bentmasreya.net/
http://www.freekareem.org/
http://hakazaana.blogspot.com/
http://www.id3m.com/index.php/
http://kefayaheroes.blogspot.com/
http://misrdigital.blogspot.com/
http://malek-x.net/
http://ma3t.blogspot.com/
http://m-monem.blogspot.com/
http://molotoofy.blogspot.com/
http://monem-press.blogspot.com/
http://www.omraneya.net/
http://sharqawy.wordpress.com/
http://www.speaksfreely.net/
http://t5at5a.blogspot.com/
http://wa7damasrya.blogspot.com/
http://www.youtube.com/AlJazeerachannel/
http://video.alarabiya.net/
http://www.youtube.com/AlJazeerachannel/

KATHARINA NÖTZOLD

Do the Different Formats of the Lebanese Media constitute a Pan-Arab Public Opinion?

Much has been written about transnational Arab satellite channels and a number of mostly Western observers have been euphoric about the channels' potential to democratise the Arab public sphere.[1] However, some caution should be applied. Since the Arab Middle East has been mostly characterised by the rule of authoritarian governments, a gradually emerging public sphere cannot be simply understood as the Habermasian ideal of an area of negotiation between civil society and the state. This would require as a minimum that non-state players were able to voice their arguments freely in public, in order to break the discursive and political monopoly of authoritarian regimes. So far, this has not been the case in most Arab countries. Marc Lynch asserts the need to look at actual public discourse in order to form an honest picture of the political potential of an Arab civil society.[2] Regarding political influence, the effects of such a public sphere are still limited, partly because it is still in its infancy. Additionally, as Egyptian analyst Magdi Khalil noted, 'the Arab street is cut off from the international street in its concerns and its goals – globalisation, the environment, human rights, unemployment, women's rights, freedom of religion, right to development.'[3] Instead, Arab news media mainly concentrate on Arab issues such as Palestine, the Iraq War and to a certain degree on reform in the Arab world.[4] Indeed, the Arab public sphere 'can mobilise public outrage, pressure leaders to act through ridicule or exposure [...], and even incite street protests. But it cannot, in and of itself, act.'[5] To sum up, with the existence of different talk shows broadcast on pan-Arab satellite television, Arabs have the opportunity to talk about the problems in the Arab world, but at the same time no opportunity to offer any solutions.[6] Consequently one could argue that Arab media serve as a catalyst for change, but I follow Haroub's argument that it would heap too much responsibility on them to give them the leading role as a 'primary driving force for reform'.[7]

As these introductory statements have shown, the relatively young field of Arab media studies has been characterised by a preoccupation with the phenomenon of al-Jazeera,[8] in particular with issues around news, and thus is mainly interested in the effects of these news channels on Arab society which is reflected in the topics of many conferences on Arab media. A smaller number of media scholars have started to take a closer look at those programmes mainly associated with entertainment but which have recently received more coverage because they have been the talk of the Arab world.[9] For that reason, the focus is mainly on Lebanese stations, because they have been at the centre of contention in the Arab world. Despite the entertaining focus of these shows, they have drawn strong reactions that go far beyond the nature of entertainment and have been attributed different traits. Some people greeted these shows as a challenge to authoritarian rule and decrepit traditions[10] that have hindered socio-political developments in the Middle East long enough. Others, such as the imam of the Grand Mosque of Mecca, viewed them as yet another sort of Western 'weapons of mass destruction that kill values and virtue'. Consequently, my objective is to analyse reasons why some Lebanese programme formats attract audiences not only in Lebanon but throughout the Arab world.

Unfortunately I am unable to give reliable audience figures for these programme formats to show how successful these programmes are by numbers alone. One reason is that audience studies are conducted by polling institutes often closely associated with the stations.[11] Audience ratings define the price for advertising and therefore each station has a special interest in being the market leader to receive the highest advertisement revenues.[12] Additionally, as a junior researcher it is very difficult to obtain information about audience figures for non-commercial purposes by companies such as IPSOS or PARC who mainly provide the relevant data to paying customers and who guard their audience figures almost like state secrets out of fear not to lose out to their competitors in the market.

The outlook of the Lebanese television landscape

Lebanon's audiovisual media landscape is characterised by its sectarian outlook, that is, Lebanese television stations reflect and reinforce the characteristics of Lebanon's political and sectarian society which finds its expression in the confessional outlook of the stations' majority shareholders. At present, seven licensed television stations are on air and all but state-run Télé Liban operate terrestrial and satellite stations.[13] Lebanese Broadcasting Corporation International (formerly LBC and known to all by that acronym) is seen as speaking for Maronite Christians.[14] Saudi prince Alwaleed bin Talal owns 49 per cent of LBC's satellite operation, LBCSAT.[15] Future TV is seen as addressing mainly Sunni Muslims and is owned by the family of the late Prime Minister Rafiq Hariri and his political and economic allies. The National Broadcasting Network (NBN) – jokingly also referred to as Nabih Berri Network – is owned by the family and political allies of Shi'i parliamentary speaker Nabih Berri whereas al-Manar serves as the organ of the Shi'i political party and militia Hizbullah. New TV is owned and run by Lebanese businessman Tahseen Khayyat,[16] foe of the late Prime Minister Rafiq Hariri and finally, Télé Liban – the supposedly national television station – is run by a board politically appointed by the government. Murr TV, which represented mainly the Greek-Orthodox community in Lebanon, was closed down for political reasons in September 2002 because it increasingly served as an opponent of the pro-Syrian forces in Lebanon.[17] After his return from exile in May 2005, his subsequent fallout with the anti-Syrian opposition group March 14 and his new alliance with Hizbullah and other pro-Syrian groups to form the opposition to the new anti-Syrian government of Fouad Siniora, former army commander General Michel Aoun felt that no Lebanese TV station represented the view of his Free Patriotic Movement. Consequently, the creation of Orange TV[18] was announced, an operation that is masterminded by Roy Hashem, Aoun's son-in-law. Since OTV started its broadcasting trial in July 2007, the station's political programmes mainly feature guests close to the opposition.[19]

Different Formats of the Lebanese Media: a Pan-Arab Public Opinion?

Many programme formats on LBC, Future TV and Murr TV, and to a lesser degree on New TV and al-Manar, were and are watched widely throughout the Arab world on the satellite channels of these Lebanese stations. Moreover, these programme formats usually introduce the general Arab public to content and forms of presentations that are novel in Arab television. They are mostly adaptations of formats created and tested successfully in Western Europe and the United States.

The programme formats of Lebanese stations that are of interest here are socio-political talk shows and entertainment programmes such as game shows variety shows and reality television. The reason for this selection is quite simple – they all attract a large pan-Arab audience, although for different reasons and in varying degrees.

The 1990s: socio-political talk shows

When LBC and MTV started to broadcast socio-political talk shows such as *al-Shaater Yehki* [The Wise One Speaks], *Hadith al-'Umr* [Dialogues of a Lifetime], *Kalam al-Nas* [People's Talk] or *al-Hakeh Baynatnah* [The Talk Between Us] and *Sajjil Mowqaf* [Take a Stand] in the 1990s, they drew crowds but they also raised the ire of some viewers because these shows soon became known for treating topics that were deemed taboo. Lebanese sociologist Melhem Chaoul conducted a study in which he scrutinised the topics of seventy-seven episodes of LBC's show *al-Shaater Yehki* from April 1995 to June 1998.[20] Regarding the subject matters, thirty-one episodes (39.7 per cent) were about beliefs, traditions and communication (committing suicide, racism), whereas eighteen episodes (23.1 per cent) dealt with economic and social life (rape, prostitution), followed by seventeen episodes (21.8 per cent) that were concerned with the private life of individuals and families (marital betrayal, domestic violence, cohabitation) and only twelve episodes (15.4 per cent) dealt with topics related to the government and public issues (revenge, bribes). Yet contrary to Chaoul's findings about the content of the episodes, in Lebanese public opinion this talk show was mainly associated with addressing individual and family problems and touching on culturally sensitive topics that were not talked about before in public, such as rape or prostitution. What was truly shocking for many in the audience was that there were actually people who spoke publicly about topics that were officially considered to be non-existent in the Arab world, and were and still are mostly associated with Western immoral behaviour. Moreover, invited talk-show guests were not only experts dealing with these issues professionally, such as psychologists or criminologists, but for the first time viewers were exposed to people who knew best about the topic from first-hand experience. It was not only the content that was daring and new to the audience; another innovation in Arab television was that viewers were actively participating through phone-in sessions and the studio audience took part in the discussions. These programme formats served as forerunners in the Arab world and influenced – along with American formats – the development of talk shows like al-Jazeera's *Akthar Min Ra'i* [More than One Opinion] or *Bila Hodoud* [Without Borders].

For almost a decade Future TV has broadcast the very popular weekly socio-political talk show *Sire wa Infatahit*. The host Zaven Kouyoumjian intends, with his topics, to attract young Arabs, discussing their everyday problems ranging from religiously mixed marriages and homosexuality to drug abuse. Although some of the topics he discusses with his guests are political at their core, his subjects usually do not challenge Arab rulers. Instead the content of his shows addresses mainly the social problems of his audience. Zaven, as he is known by his audience, also has his own website so that viewers can share their views on the show's themes. Depending on the show's subject, people from all over the Arab world not only call in during the live broadcast, but they use the internet for sharing their thoughts and often grievances. Going by the entries on Zaven's website and judging from live phone-ins during the show, it is safe to say that his programme is widely watched in Lebanon and by residents of the Arab Gulf states.

Entertainment programmes: variety shows

With the spread of satellite dishes starting in the mid-1990s, LBCSAT and Future SAT, followed later by Murr TV, targeted audiences in the Arab Gulf with a lucrative advertising market to cover production costs. With the specific targeting of the pan-Arab market, Lebanese television stations took a mixed approach. Their entertainment programmes were and still are by all means daring, colourful and trend-setting. This certainly attracted a large entertainment-starved audience, especially in Saudi Arabia where only in autumn 2006 was the first feature film released, and where entertainment activities outside private homes are limited to non-existent. At the same time, this meant TV producers had to walk a thin line, as for a largely conservative audience some of the content and the outlook of entertainment programmes were and still are shocking, immoral and should be prohibited.

Lebanese stations produce numerous variety shows from modelling contests and game shows, which are embedded in a bouquet full of singing and dancing. The most successful Friday evening variety programme is surely LBC's long-running *Ya Lail Ya 'Ain* [these are the traditional opening words of a ballad, literally 'Oh, Night, Oh, Mine Eye'], showing young men and women dancing together, guessing the tune, singing and 'setting the world on fire' as the show's promotion claims. What makes it even more attractive to viewers is that the show plays on the battle of the sexes, young men vs. young women, the latter often wearing revealing clothes. The sexual tension and the revealing clothes are considered by many in this business as elements that viewers especially in the conservative Gulf States find appealing, which ensures high audience ratings.[21]

Entertainment programmes: reality shows

What really made headlines in the Arab world, and especially among Arab youth, were the successful reality programme formats mainly introduced by Lebanese stations LBC and Future TV, namely *Star Academy* and *Superstar*. Yet, how new was reality television to audiences in the Arab world? I follow Joe Khalil's line of argument,[22] and his evaluations are identical to my observations of Lebanese television. There existed several programme formats prior to the boom of *Superstar* and *Star Academy* that included elements of reality television (mainly the competitive aspect) and which served as useful testing grounds for the big shows that were to follow. These shows were not the twenty-four-hour programmes commonly associated with reality television, but Arabic versions of Western formatted game shows. Whereas Murr TV started as early as 1997 with an adaptation of *Family Feud*, LBC started a little later to produce *Fort Boyard*, a French programme format, and MBC produced the highly successful *Who Wants to be a Millionaire?* The success of these shows encouraged producers to buy further international programme formats and at the same time they trained technical teams, and used producing these shows as practice, promoting them as 'local'. Additionally, a high level of interest from the audience and advertising clients alike convinced the management and advertising executives of television stations to produce more of these shows.

The *Miss Lebanon* contests of 2002 and 2003 served as a training ground for LBC's core staff, which later produced *Star Academy*. The finalists of the beauty contest lived together in The House, which in fact was a luxurious villa and spa. The public – which not only consisted of Lebanese viewers but included pan-Arab audiences because *Miss Lebanon* was also broadcast on LBCSAT – voted for the removal of two candidates in a weekly performance show. This happened via short text-messaging from cell phones – introducing a new tool of interactivity. Meanwhile, the contestants were being monitored by cameras and their lives and activities, such as lessons in how to dine in an expensive restaurant or the girls' workouts, were transmitted on a twenty-four-hour channel. The monitoring was especially utilised as testing ground for the set of rules of conduct in respecting 'Arab values' while cameras filmed the contestants around the clock.

Before *Superstar*'s launch in January 2003 on Future TV, a carefully orchestrated advertising campaign started to create interest among the prospective audiences for a new kind of show that seemed eagerly anticipated. Sure enough, it became the talk of the town immediately after first going on air. I overheard numerous conversations between various people in Beirut's shops, buses and service taxis, discussing the participants and their chances. Although *Superstar* was not, in the strictest sense, a reality format but more a variety show, it contained elements of reality shows by showing the daily activities of the contestants in short segments. *Superstar*, the licensed Arab version of Fremantle Media's British programme *Pop Idol*, was intended to be a transparent pan-Arab search for the next superstar singer, using a panel of judges headed by Elias Rahbani, casting calls, live performances and the audience's right to vote. The *Pop Idol* format was successfully sold to other countries worldwide but its Arab version was the first one featuring participants from an entire region, and casting drives were held in different parts of the Arab world. There were certainly some differences to spot between participants from the various Arab countries – different

Arabic dialects spoken, different fashion statements made – especially at the initial casting sessions[23] – and sometimes even different Arab musical traditions cast during the first sessions. *Superstar* managed to create a united Arab audience watching the show from Morocco to Iraq, and around thirty million viewers watched the finale of *Superstar 1*.[24] However, the audience split along nationalist lines when voting for the superstar – much like the Eurovision song contest does – even though this was not intentionally set up by the show's producers. The massive voting campaign that involved Jordan's King Abdullah and Jordanian mobile phone companies to ensure that Jordanians voted for the Jordanian candidate became legendary.[25] Even before the finale, a riot had broken out when Lebanese fans thought they saw a conspiracy by the Syrian secret service, putting pressure on Future TV's executives to eliminate Lebanese candidate Melhem Zein during the vote, whereas Syrian candidate Rowayda Attieh was elevated to the finale. In fact, political tensions in Lebanon ran so high that the Lebanese believed the Syrian secret service capable of influencing even trivial non-political decisions.[26] Elias Rahbani saw fault in Zein's statement that he is Arab and not Lebanese, which alienated many of the Lebanese from voting for Zein.[27]

The by far most successful Arab reality show *Star Academy* started broadcasting in December 2003 on LBC. Here again, a Western (French) format was adapted to the Arab market. Young Arabs were invited to join the academy, where they lived together and trained to become singers. After initial courses for all participants, the instructors nominated two candidates for the weekly prime episodes and the pan-Arab public would vote by phone for one of them to stay in the academy. This again was all covered twenty-four hours a day on LBC-owned music station *Nagham*, which was specially dedicated to *Star Academy*. As a tribute to Arab cultural sensitivities, participants of the show were not allowed near the bedrooms of the opposite sex. The places where they spent time together were the classrooms, the living room and kitchen. However, this free mingling of young people of both genders unrelated to one another but living together under one roof raised the ire of many conservatives in the Arab world.[28] Famous al-Jazeera preacher Yussuf al-Qaradawi complained that 'reality TV and Star Academy especially are instruments of cultural and intellectual invasion of the ummah'[29] because these programmes distract the

Muslim youth from their religion and from real political issues. A popular Saudi campaign denounced the show as *Akademiyyat al-Sheitan* (Devil's Academy) and it urged for boycotts of such shows.[30]

Reality shows such as *Star Academy* or *al-Wadi* [The Valley][31] – the latter featuring famous Lebanese bosom wonder Haifa Wehbi who needs help with her farm – are condemned by clerics in the Arab world, whereas some observers in the Western world were enthusiastic that these shows would spread democracy and Western values in the Arab world because people vote for their favourite candidates. More on the sidelines, the reality format was equally condemned by conservatives in the Western world. Reality TV will not bring democracy to the Arab world just because people vote for their favorite candidate but it has certainly managed to grab the attention of many people and has received an extraordinary amount of commentary in TV shows like al-Jazeera's *al-Itijah al-Muaqis* [The Opposite Direction] or LBC's *Kalam al-Nas* [People's Talk], and many more op-eds discussing the effects of these shows have been published in newspapers across the Arab world. It is true that reality shows use audience participation and voting to determine the winners of these competitions and that reality television introduced this feature to countries that are not known for holding regular parliamentary or presidential elections in which people have more than one choice. Yet the events on the ground speak a language of their own: young Saudis voted for their compatriot Hisham who won *Star Academy 2* despite strong opposition from the clergy and an official ban on voting on *Star Academy* by the main Saudi mobile phone provider, but at the same time young Saudis did not show much interest in participating in the municipal elections. However, what gives these shows the air of being somehow political and what makes them successful is that they seem to be a rather fair competition where participants are not chosen because of the right connections. What counts are their own creativity, skills and personal initiative. This 'reality' of the reality show is different from experiences many young Arabs have in their daily lives, where they are prevented from expressing their opinions and where they only get jobs because of *wasta*[32] and rarely because of competence.[33]

Proponents of pan-Arab unity may be dissatisfied that although these programmes are targeting a pan-Arab audience, they do not create automatically pan-Arab unity among the audience but instead, when it comes to voting, the quality or merit alone of the candidate no longer counts, just the candidate's nationality. It seems to reinforce particular national identities instead of one pan-Arab identity. In the end this means that these shows create different discourses in each Arab state. A case in point is when Palestinian singer Ammar Hassan had to concede to the Libyan contestant in *Superstar 2*. Palestinians had appealed to Arab solidarity with Palestinians and their struggle by supporting their candidate and they were bitterly disappointed by what they saw as lack of support for their cause.[34]

Other, non-Lebanese, stations have tried also to use the reality TV format for their purposes. The reality matchmaking show *al-Hawa Sawa* [Living Together] transmitted by one of ART's twenty-four-hour channels featured young women courting a marriage proposal on air. This show functioned as bait to test how conservative audiences would react to the reality format on a station that belongs to Saudi businessman Sheikh Saleh Kamel, who also sponsors the religious channel *Iqra*. The show was completely adapted to consider 'Arab values', meaning that women had to adhere to strict dress codes and other rules that governed their personal conduct (tattoos and smoking were banned). One participant left the show because she was bored and the winner refused to marry her designated groom and no other reality projects followed at ART. In early 2004 MBC ran the show *al-Ra'is* [The Boss, The President] which is known in the Western world as the Endemol production *Big Brother*. While effort was made to adapt the show to accommodate religious and cultural sensitivities, *al-Ra'is* nevertheless caused uproar in Bahrain where the Big Brother house was situated and was cancelled after some episodes. It was the first and last attempt of producing a twenty-four-hour reality show outside Lebanon and it exposed the divisive nature of reality television along conservative religious versus liberal lines.

Conclusion

In the end I am reluctant to say that Lebanese programme formats have the ability per se to constitute a pan-Arab public opinion. Although these shows are intended for a pan-Arab audience, the stories are legion that the majority of people vote for their participating compatriots instead of honouring merit and performance. The pan-Arab aspect of these shows is that usually the entire Arab world is involved in a single, common action and public dialogue about the same issue,[35] and this could be mistakenly viewed as contributing to an Arab public sphere by itself. But the discourses that they raise are specific national discourses of and within the different Arab countries, focusing on the domestic public. This was underlined not only by voting patterns of the audience or politicians who supported their compatriots but also by contestants themselves who raised the flags of their nations, some of whom made it clear in interviews that they were sons or daughters of a specific country. Moreover, calls in religious circles to ban these shows were led mostly in the more conservative Gulf countries, whereas in Lebanon intellectuals complained rather about the mindlessness of these shows, which highlights again a difference in discourse among different Arab regions.

It can be safely said that Lebanese programmes have paved the way for new trends in Arab television, such as the introduction of reality television formats and interactive voting via mobile phones, thus tapping a large source of income for television producers[36] and influencing other Arab stations to imitate the Lebanese stations. This is most certainly the result of Lebanon's traditional role of looking westward for inspiration. Lebanese TV stations, especially LBC and MTV, started early on to cooperate with American and French television stations. When LBC and MTV aired their first socio-political talk shows in the 1990s, they started to challenge existing taboos in Lebanese society and thus tested the ground for their future endeavours on the pan-Arab market. In contrast to most Western producers, who are usually not familiar with the intricacies of Arab society, the Lebanese producers of these shows are aware of them and may play with them in such a way that they break taboos. However, they still get by without having to cancel any of these shows because they do this from Lebanon, where society generally has a reputation among Arabs as being morally less rigid. Lebanese producers successfully adapt Western programme formats to Arabised versions of the original, with a distinct Lebanese flavour. This flavour is a mix of using young, beautiful presenters wearing revealing clothes and speaking in a colloquial language which young people find appealing, interspersed with French and English phrases which add a cosmopolitan flavour to it. Within different cultural traditions in the Arab world, only reality shows that were produced in Lebanon were commercially successful. Non-Lebanese channels that wanted to ride on the same wave of success but adhered to stricter 'moral' rules targeting more conservative audiences in the Gulf have so far been unable to gain the same size of audience. Having said that, it is no surprise that despite some protests by conservatives, young people of both sexes could live together in *Star Academy* broadcast from Lebanon but not in *al-Ra'is* which was broadcast from Bahrain by a non-Lebanese station.

Besides the economic gains, Battah describes how with SMS text-messaging popularised by *Star Academy*, communication barriers upheld by conservatives were broken down by young people sending flirtatious messages to each other, and despite the religious ban to vote for *Star Academy* young Saudis could outwit religious police by using a multitude of websites offering SMS services.[37] Some variety and reality programme formats certainly attract a large audience, especially young people, and they often imitate styles set in these programmes which could be interpreted in the way as *Superstar* producer Christine Jammal did, that such a show 'is uniting the Arab world.'[38] However, this would be too optimistic because even though young Arabs share common experiences, they face different social problems in their respective countries and have to deal with them in their specific environments.

Notes

1. Louay Y. Bahry, 'The New Arab Media Phenomenon: Qatar's Al-Jazeera', *Middle East Policy* VIII, no. 2, 2001, pp. 88–99; Bradley, Ed., '60 Minutes: Inside Al-Jazeera.', CBS, 2001, [www.cbsnews.com/new/story/0,1597,314278-412,00.shtml (accessed 15 December 2001)]; Thomas Friedman, 'Glasnost in the Gulf', *New York Times*, 27 February 2001; Thomas Friedman, 'Fathers and Sons', *New York Times*, 12 February 1999: A 27; Gary C. Gambill, 'Qatar's Al-Jazeera TV: The Power of Free Speech', *Middle East Intelligence Bulletin*, no. 2, issue 5, 2000, [www.meib.org/articles/0006_me2.htm (accessed 30 December 2001)]; Edmund Ghareeb, 'Revolution in the Arab World: An Assessment', *Middle East Journal*, no. 54, issue 3, 2000, pp. 395–418.
2. Marc Lynch, *Voices of the New Arab Public*, New York, 2006.
3. Magdi Khalil, 'The Opposite Direction', *al-Jazeera*, 16 April 2002; quoted in Lynch, p. 79.

4. Lynch, pp. 79–83.
5. Ibid., p. 54.
6. Khaled Haroub, 'Sociology of the Arab Media', *Wajhat* (UAE), 10 January 2005.
7. Lynch, p. 54.
8. Walter Armbrust, 'Letter from the editor: Al-Jazeera is not a medium', *Transnational Broadcasting Studies*, no. 2, 2005, p. 1.
9. See, for example, Joe Khalil, 'Blending in: Arab Television and the Search for Programming Ideas', *Transbroadcasting Journal*, no. 13, 2004 [http://www.tbsjournal.com/Archives/Fall04/khalil.html (accessed 5 November 2006)]; Joe Khalil, 'Inside Arab Reality Television: Development, Definitions and Demystification', *Transbroadcasting Journal*, no. 15, 2005 [http://www.tbsjournal.com/Archives/Fall05/Khalil.html (accessed 12 November 2006)]; Marc Lynch, *Voices of the New Arab Public*, New York, 2006; Marwan Kraidy, 'Popular Culture as a Political Barometer: Lebanese-Syrian Relations and Superstar', *Transbroadcasting Journal*, no. 16 2006 [http://www.tbsjournal.com/Kraidy.html (accessed 5 November 2006)]; Kraidy Marwan, 'Idioms of Contention: *Star Academy* in Lebanon and Kuwait', in N. Sakr (ed.), *Arab Media and Political Renewal. Community, Legitimacy and Public Life,* London: 2007, pp. 44–55; Imad Karam, 'Satellite Television: A Breathing Space for Arab Youth?', in N. Sakr (ed.), *Arab Media and Political Renewal. Community, Legitimacy and Public Life,* London: 2007, pp. 80–95.
10. For example Thomas Friedman of the New York Times but also other conservative US-American writers have mused about the democratising character of these shows.
11. Lebanese Broadcasting Corporation (LBC) gets its audience ratings from IPSOS which is an international organisation conducting audience and market research. However, the Lebanese branch of IPSOS has ties with Antoine Choueiri who is the main advertising contractor for LBC. As a result, LBC's main competitor Future TV which also used to receive their ratings from IPSOS, did not trust the objectivity of IPSOS and managers at Future claimed that IPSOS' numbers were always biased in favor of LBC.
12. Future decided to have their own market research company and as a result their ratings were at times better than LBC's because they used a different sample of people. Bernard Menhem of FMS (Beirut), interview by author, 2 May 2003. He explained that IPSOS mainly asked Lebanese living in the Mt. Lebanon region which is mainly inhabited by Lebanese Christians who in their majority watch LBC as their first choice. According to Menhem, Future TV is mostly watched by Sunnis who live in Beirut and in and around Sidon, regions that were not covered by IPSOS with

13. the same intensity as the Mt Lebanon area. Therefore Menhem and other Future TV staff interviewed by the author claimed that audience figures were not truly representative of the situation on the ground.
13. Télé Liban only operates terrestrial broadcasting despite government plans to transmit also via satellite. The station is chronically under-funded and over-staffed and each new government promises to get TL out of its economic dire straits. However, no government seriously attempted to reform TL's operations.
14. It was founded in 1985 by the Lebanese Forces (LF) but several changes of policies by majority shareholder and CEO Sheikh Pierre Daher since LBC's establishment were intended to publicly distance LBC from the LF in order to relieve LBC of political pressure, especially after the imprisonment of LF commander Samir Geagea in 1994. This did not deter most Lebanese to associate LBC continuously with the LF at least in regards to its political point of view in Lebanese news.
15. 'Saudi's Prince Alwaleed buys 49 per cent stake in LBCSAT', *Arab Ad*, December 2003, p. 152.
16. Tahseen Khayyat is a Sunni businessman hailing from the same southern city of Sidon as Rafiq Hariri. More recently, New TV has become the station of the opposition to the current government of Prime Minister Fouad Siniora, attracting for its news especially also Christian viewers who are allied with General Michel Aoun.
17. Although this decision was revoked by Lebanese parliament in August 2005, MTV as it is known in Lebanon has not managed to reopen in the beginning of 2006 as was promised and it remains to be seen whether it will be on the air again any time soon.
18. Orange is the party colour of Aoun's Free Patriotic Movement.
19. OTV staff underlined that they try hard to invite representatives of the Siniora government to avoid being blamed as biased towards Aoun but the government representatives would boycott the station. Author's email interview, 16 September 2007.
20. Melhem Chaoul, 'The Face of Society: Samples of Social Files', *Bahithat* VI, 1999–2000, pp. 290–316 (in Arabic).
21. Joe Khalil, 'Blending in: Arab Television and the Search for Programming Ideas', *Transbroadcasting Journal*, no. 13, 2004 [http://www.tbsjournal.com/Archives/Fall04/khalil.html (accessed 5 November 2006)].
22. Joe Khalil, 'Inside Arab Reality Television: Development, Definitions and Demystification', *Transbroadcasting Journal*, no. 15, 2005 [http://www.tbsjournal.com/Archives/Fall05/Khalil.html (accessed 12 November 2006)].
23. Some participants from Gulf countries wore traditional attire, wearing the thawb.

24. Superstar 2 had 75,000 hopeful applicants from twenty-two states. One casting took also place in Paris to include Arab immigrants. Hannes Alpen, Die TV-Show 'Superstar' demokratisiert den Nahen Osten', *Zenith*, no. 1, 2004, p. 34.
25. During the author's visit to Jordan in November 2003 all the Jordanians voiced their pride that the winner Diana Karazon originated from their country and that Jordan for once had won in an Arab competition.
26. Marwan Kraidy, 'Popular Culture as a Political Barometer: Lebanese-Syrian Relations and Superstar', *Transbroadcasting Journal*, no. 16, 2006 [http://www.tbsjournal.com/Kraidy.html (accessed 5 November 2006)].
27. Elias Rahbani, Liebe ohne Grenzen', interview by Hannes Alpen, *Zenith*, no. 1, 2004, p. 35.
28. Students of the Academy hugged and kissed but after living and studying together for almost four months, they had become like brothers and sisters and several articles mocked religious conservatives because young people hug and kiss also outside television. See for example Hania Taan, 'In praise and in sorrow at the passing of Star Academy', *Daily Star*, 3 April 2004.
29. Ali Mohammad Taha, 'Qaradawi: Reality television inflames the youth of the umma and is an instrument for cultural invasion', *Asharq Al Awsat*, 28 April 2004.
30. Faisal Abbas, 'Satan's Academy', *Asharq Al Awsat*, 8 June 2005.
31. Al-Wadi is the Lebanised or Arabised version of the original programme format 'The Farm'.
32. This term is widely used in the Arab world and can be best translated into English as 'clout, connections or influence'. Wasta involves using of one's connections and influence in places of power to get things done outside of the normal procedures and is a form of corruption.
33. Marwan Kraidy, Reality Television and Politics in the Arab World: Preliminary Observations , *Transbroadcasting Journal*, no. 15, 2005 [http://www.tbsjournal.com/Archives/Fall05/Kraidy.html (accessed 12 November 2006)].
34. An account of Palestinian refugee camp Ain al-Hilweh near the Lebanese city Sidon sums up the different discussions. Mahmoud Assadi, public relations official with the PLO, said the organisation has decided to settle the costs of all calls via internet in Ain al-Hilweh even though not all inhabitants of the camp agreed with the preparation for the finale to support the Palestinian candidate. One vegetable shop owner argued that the Palestinian people were drowning in blood baths while refugees danced and sang and he accused Israel for distracting Arabs from fighting for Palestine whereas a butcher stressed that the issue was part of the resistance against Israel and victory for the Palestinians. Additionally, Palestinians asked Shiites in South Lebanon to vote for the Palestinians

reminding them that Libya was responsible for the disappearance of Shiite leader Moussa Sadr in the 1970s. Mohammad Zaatari, 'Ain al-Hilweh leads push for Palestinian Super Star' ', *Daily Star*, 20 August 2004.
35. Rabih Khoury, 'Star Academy Brings People Together', *Daily Star*, 29 March 2004; Mayssam Zaaroura, 'Reality television the new gospel or a passing trend?', *Daily Star*, 21 January 2004.
36. In terms of telephone calls alone, industry sources say LBCI generated over $3 million per week throughout the duration of the sixteen-week show. Habib Battah, ,Star Academy makes TV history. Popular reality show brings in unprecedented revenues', *Daily Star*, 2 March 2004.
37. Habib Battah, 'SMS: The Next TV Revolution', *Transbroadcasting Journal*, no. 16, 2006 [http://www.tbsjournal.com/Battah.html (accessed November 5, 2006)].
38. Hannes Alpen, 'Die TV-Show "Superstar" demokratisiert den Nahen Osten', *Zenith*, no. 1, 2004, p. 34.

Bibliography

Alpen, Hannes, 'Die TV-Show "Superstar" demokratisiert den Nahen Osten', *Zenith*, no. 1, 2004, p. 34.

Armbrust, Walter, 'Letter from the editor: Al-Jazeera is not a medium', *Transnational Broadcasting Studies*, 2005, no. 2, p. 1.

Bahry, Louay Y., 'The New Arab Media Phenomenon: Qatar's Al-Jazeera', *Middle East Policy VIII*, 2001, no. 2, pp. 88–99.

Battah, Habib, 'SMS: The Next TV Revolution', *Transbroadcasting Journal*, no. 16, 2006 [http://www.tbsjournal.com/Battah.html (accessed 5 November 2006)].

Bradley, Ed., '60 Minutes: Inside Al-Jazeera', *CBS*, 2001, [www.cbsnews.com/new/story/0,1597,314278-412,00.shtml (accessed 15 December 2001)].

Chaoul, Melhem, 'The Face of Society: Samples of Social Files', *Bahithat* VI, 1999–2000, pp. 290–316 (in Arabic).

Gambill, Gary C., 'Qatar's Al-Jazeera TV: The Power of Free Speech', *Middle East Intelligence Bulletin*, 2000, no. 2 (5) [www.meib.org/articles/0006_me2.htm (accessed 30 December 2001)].

Ghareeb, Edmund, 'Revolution in the Arab World: An Assessment', *Middle East Journal*, 2000, no. 54 (3), pp. 395–418.

Haroub, Khaled, 'Sociology of the Arab Media', *Wajhat* (UAE), 10 January 2005.

Karam, Imad, 'Satellite Television: A Breathing Space for Arab Youth?', in Naomi Sakr (ed.), *Arab Media and Political Renewal: Community, Legitimacy and Public Life*, London: I. B.Tauris, 2007, pp. 80–95.

Khalil, Joe, 'Blending in: Arab Television and the Search for Programming Ideas',

Transbroadcasting Journal, 2004, no. 13 [http://www.tbsjournal.com/Archives/Fall04/khalil.html (accessed 5 November 2006)].

Khalil, Joe, 'Inside Arab Reality Television: Development, Definitions and Demystification', *Transbroadcasting Journal*, 2005, no. 15 [http://www.tbsjournal.com/Archives/Fall05/Khalil.html (accessed 12 November 2006)].

Kraidy, Marwan, 'Reality Television and Politics in the Arab World: Preliminary Observations', *Transbroadcasting Journal*, 2005, no. 15 [ttp://www.tbsjournal.com/Archives/Fall05/Kraidy.html (accessed 12 November 2006)].

Kraidy, Marwan, 'Popular Culture as a Political Barometer: Lebanese-Syrian Relations and Superstar', *Transbroadcasting Journal*, 2006, no. 16, [http://www.tbsjournal.com/Kraidy.html (accessed 5 November 2006)].

Kraidy, Marwan, 'Idioms of Contention: *Star Academy* in Lebanon and Kuwait', in Naomi Sakr (ed.), *Arab Media and Political Renewal: Community, Legitimacy and Public Life*, London: I. B. Tauris, 2007, pp. 44–55.

Lynch, Marc, *Voices of the New Arab Public*, New York: Columbia University Press, 2006.

YASSIN MUSHARBASH

In the 'Net' of Public Opinion – Islamist Online Media at Work

During the last couple of years experts all over the world have noticed a clear trend: ever more Islamist, Jihadist and terrorist organisations are making extensive and creative use of the Internet. By now, this development has also become an issue of concern in the security echelons as well as in political quarters across the world, as well as for the media, which devote more resources to covering these online activities than before. In this article, I will attempt to shed some light on the following questions: What are Islamist online media? To what extend do they interact with public opinion? What does it mean to see them 'at work'? Is there such a thing as a dialogue between al-Qaʿida and the West, i.e. a virtual dialogue with the enemy?

What are Islamist online media?

Generally, there are two kinds of online media that qualify to be discussed in this context: organised and unorganised media or, more to the point, those attached to and directed by certain organisations or movements, such as the website of Hizbullah, and those who are not attached but are rather run by independent supporters of Islamism or Jihadism. The main difference seems to be that for the most part moderate Islamist movements run their own websites or other online media, while the more radical and violent have by now almost completely taken to making use of online outlets established by independent supporters to get their message across.

In terms of different formats, it also seems fair to say that the more radical organisations are more active and more creative. As examples will show, al-Qaʻida, for one, and similar networks are very active publishers – online magazines are just one example, albeit an important one.

This opens up the logical question of who runs these media, and who consumes them? While in the case of official websites it is clear that the publishers are cadres of the respective organisations, matters are more complex in the case of independent and especially jihadi media. There are certain magazines that can clearly be traced back to the core of al-Qaʻida, for example, but more and more independent freelancers are taking over. They have even taken over formerly official brands of al-Qaʻida, like the Global Islamic Media Front, once a news outlet for Bin Ladin and now a network of freelance media activists with no exclusive relation to al-Qaʻida any more. Still, the GIMF is very influential, as will be shown.

As for the consumers, we have no clear picture yet but assumptions can be made based on what we can read on websites, discussion forums and message boards. The readership seems to be mainly young, male Arabs – to what extent they live in the West and in the East is uncertain. Nevertheless, the clear trend to supply sympathisers with translations of the originally Arabic material, points to a large segment among second or third generation Muslims in the diaspora.

To what extent do these media interact with public opinion?

There is a very clear sense among Islamists who spend a lot of time online that the Internet is a valuable tool, especially because of the high degree of media control and the suppression of oppositional voices in most Arabic countries. In this sense – and in this sense only – one could say that the sum of what we can read on Islamist online media is a form of public opinion in itself. It is by far the best way to find out what Islamists think. Debates that have been held there, across continents and time zones, have shown this clearly, e.g. the discussion about the legitimacy of the London underground bombings of 2005 and the 2007 summer war in Lebanon. In both instances, discussion was fierce. In the London case, the more radical side seems to have had the upper hand in the end. In the Lebanon debate, the main issue was whether it was acceptable to help or even wish for Hizbullah's success. This was mainly denied with the argument that the Shi'i party was a movement of unbelievers (*hizb al-lat*!).

At the same time, there is interaction between this 'public' and the mainstream of international public opinion. Islamist online media like the GIMF, as will be pointed out in detail, are taking a two-way approach: they want to unite the Muslim and the Islamist opinion and they want to influence Western opinion. In order to achieve this they even distance themselves from al-Qa'ida and stress that they do not act in the name of any particular organisation. Rather, they suggest they are speaking for all Muslims.

This strategy leads to the presentation of interfaces for interaction. Important jihadi Internet forums have started to issue accreditations for journalists of Western media, who they willingly supply with communiqués from al-Qa'ida and other jihadi groups.

What does it mean to see them 'at work'?

Islamist online media follow multiple purposes. It seems fair to say that the more moderate an Islamist organisation is, the more its online outlets serve to inform its followers. On the other hand, the more radical and violent an organisation is, the more its media outlets serve to instigate people to act. One example is the multiple ways in which al-Qa'ida and cooperating networks have made use of the Internet to inflame Muslims over the Danish cartoons. The Ansar network published the names of European newspapers that reprinted some of the pictures and a Pakistani student in Germany tried to stab the editor-in-chief of a German daily that had been named. There was probably no direct link, but what happened was exactly what the media activists had intended.

Another very important point in this context is the way Islamist organisations use online media to counter the mainstream media or the state-run press in their home countries. An illustration of this could be observed in 2003 when the Saudi branch of al-Qa'ida published communiqués in real time about the kidnapping and killing of a group of Westerners in their hands, just to prove that the Saudi newspapers and the Ministry of Interior were lying.[1]

Both Western and Eastern news outlets now look much closer for Islamist online media and quote them much more than some years ago. This has opened up new and as-yet unexplored ways in which Islamist online media are now making an impact even upon people who do not look for them, because they are confronted with what they publish in an indirect way and through the reporting of the mainstream media. An example of this would be the spreading of sniping videos by the 'Islamic Army in Iraq', which first appeared on their website, then on jihadi discussion forums, then on YouTube. From there, the *New York Times* took up the trail and reported about Juba, the master-sniper. The paper also published what some of those who distributed the video clips had said: namely that they felt they were passing on information that Western media would not report on. Some of these people were American citizens and not even Muslims. In the end, the Islamic Army managed to get a message across, even to the Western public.

Is there such a thing as a dialogue between al-Qai'da and the West?

As someone who has followed the mainly Internet-based activities of al-Qa'ida in the field of public relations and propaganda, I have to conclude that, yes, there is a dialogue. Even though al-Qa'ida goes a long way to denounce the West as its prime enemy, the network in fact does seek to influence this designated enemy by what it publishes. Al-Qa'ida wants the West to react – and the West does react. In the widest sense of the word, this qualifies as a dialogue, even though it might be a rather virtual one and despite the fact that this dialogue goes one way most of the time.

In fact, one might say that al-Qa'ida follows a rather sophisticated communication and information strategy in regard to the West. It has been expressed on several occasions. Based on the material al-Qa'ida does publish, the network's objectives can be identified fairly clearly: to spread ideology and know-how and win sympathisers; to create and maintain a degree of credibility; and to mislead the West and spread fear.[2]

When discussing the strategy aimed at meeting these needs, one also has to bear in mind that al-Qaʿida is no longer the same organisation it used to be in 2001. Today, we are not talking about a military cadre-organisation any more, but rather about a movement or an idea. Accordingly, al-Qaʿida has changed its behaviour. For one thing, a good deal of what al-Qaʿida publishes today is aimed at winning new recruits and supporters. For another, al-Qaʿida has opened up to its supporters, meaning that today a lot of people who were never in Afghanistan play a big role in implementing al-Qaʿidaʾs information and communication strategy.

Today, information that used to be kept secret, like al-Qaʿidaʾs training manuals, are being published on the Internet by al-Qaʿida itself – in this case as a substitute for the lost training camps of Afghanistan and a means of attracting sympathisers and recruits. These sympathisers, by the same token, then start to improve this material – which has already resulted in a wide spread of ever more modern and more dangerous bomb-building instructions to be found on the Internet.[3]

If one looks at the means al-Qaʿida chooses to implement this strategy, it shows that the network mainly uses three different approaches: messages directed solely towards the West; messages directed solely at the 'own audience'; and mixed messages.

Examples for the first kind are:

1. Bin Ladin's so-called 'peace initiative' of April 2004, in which he offered to stop attacks against Europe if these countries would withdraw from Iraq.
2. A statement hidden in a claim of responsibility by the Saudi al-Qaʿida branch in May 2004, when they explained that they killed a German citizen because Germany was a prime ally of the US, even if not engaged in Iraq.[4]

Examples of the second kind of message would be:

1. Al-Qaʿidaʾs online magazines, like *Sawt al-Jihad* and *Maʿaskar al-Battar* which are concerned with the spread of ideology and know-how. Already the cover could illustrate this. These magazines are tailor-made for young Arab Islamists and even go as far as giving e-mail-addresses where to send plans for terror attacks.[5]
2. 'Al-Qaʿida TV', a weekly newscast which started in late September 2007,

consisting of a digest of topics important to an al-Qa'ida-sympathiser, including news from the Iraqi battlefields or the leadership. The given screenshot may serve to illustrate this.

3. Another, probably the best example, is the GIMF, or Global Islamic Media Front or *al-Jabha al-i'lamiyya al-islamiyya al-'alamiyya*. Until the beginning of 2005, this was the official al-Qa'ida news agency, and material by the GIMF could be considered authentic. By autumn 2005, this organisation had completely changed hands. Sympathisers – who may or may not have connections to the fighting core of al-Qa'ida – are now running it. The GIMF has a logo and even different departments. In October 2005 the secretary general announced that they consider themselves the 'link between the mujahedin and the sons of Islam and those non-Muslims who are no enemies of Islam.' The 'Americans consider the Internet as their property' it went on. The GIMF itself, they claimed, 'belongs to no one. It is the property of the Muslims, without geographical borders.'[6] Arabic media specialists were asked to join the effort, with the aim that all al-Qa'ida and associated Internet activity should be conducted under the banner of the GIMF. This development was very important because the GIMF also plays a role as a source for Western media. This will be discussed briefly later.

4. Calls to delude the West by starting fake organisations and publishing empty threats also fall under this category. To illustrate this, one can state that in 2004, the Saudi al-Qa'ida branch called on its followers to 'found new groups with new names to confuse the West and its secret services and media'. The call was largely heeded and the confusion was great indeed.

Finally, here are some examples of the third kind, which I call 'mixed messages'.

1. A very good example is obviously Bin Ladin's speeches, which always have a double meaning – one for the 'own audience', and one for the West in the form of threats. One speech from December 2004 serves very well. It contained twenty pages in which Bin Ladin dealt with Saudi interior politics in great detail on the one hand, and three threats against the West on the other. Attention was thus assured in

both worlds, albeit the Western media missed the part about Saudi Arabia.

2. A more current example is from December 2007, when Aiman al-Zawahiri pledged on the Internet that he would accept questions from both Western journalists and al-Qa'ida supporters, and provide answers some time in the future. Thus far (March 2008) the answers have not been published.[7]

3. Also interesting are claims of responsibility for certain attacks. It is here that you get the best sense of how intensely al-Qa'ida tries to enhance its credibility. The reason is simple: an attack is worthless if no one knows who did it. The following example demonstrates how this works: in late 2005, Abu Musab al-Zarqawi, who until the American army killed him in 2007, headed the Iraqi branch of al-Qa'ida, released about a dozen claims for attacks every day. They were all published on one website first, and they all appeared with one signature – that of one 'Abu Maysara al-Iraqi', an identity that was meant to show the audience that the sender is always the same. Other organisations and al-Qa'ida branches have followed suit and operate with recognisable letterheads, etc. In this case also, the message is addressed to both the West and al-Qa'ida's own audience, because credibility must not be lost with either group.

The reaction of the Western media

An examination of the ways in which Western or, more precisely, European media react to this 'dialogue initiative' reveals certain patterns. First, it has to be noted that most Western or European journalists largely miss the way in which al-Qa'ida behaves today. Often, al-Qa'ida is still portrayed as a bottom-down network of dumb, fanatical, amoral people who have no values other than killing other people. The role that religion plays for these people is hardly ever recognised; similarly, the relationship between the large crowd of al-Qa'ida-sympathisers and their outfall with their Arab undemocratic regimes is neglected. The European media have thus only a half-picture of al-Qa'ida or Osama Bin Ladin. It is only slowly that journalists started to realise that al-Qaʻida is in fact a much more complex organisation than they had thought.

The reason for this new understanding beginning to spread is – interestingly enough – mainly due to the undisputedly near-professional communication and information strategy that al-Qaʻida has developed. This can be seen as a success for al-Qaʻida, bizarre as it sounds. One could even say that the organisation itself is changing the image we have of it, more than anything else, by becoming better at explaining what it wants and by addressing the West explicitly. Even though not all of what al-Qaʻida sends is being understood, a certain effect can be observed.

The single most important underlying reason for the confused picture most European journalists have of al-Qaʻida though is – in my opinion – the language barrier. In Germany, there is no more than a handful of journalists who speak and read Arabic and actually work on al-Qaʻida. But al-Qaʻida usually doesn't speak English – although on one occasion a Bin Ladin speech complete with English and German subtitles was sent to al-Jazeera, which again shows that al-Qaʻida is in fact sending messages to the West. Also, over the past years, more and more English al-Qaʻida videos have been produced, mostly featuring 'Adam the American', who appeals to his fellow-Americans in their own language.

However, one must of course not forget that al-Qaʻida, from a professional journalistic perspective, is not just any communication partner, but an organisation most people view as terrorist. Reporting on it poses ethical questions: do you really want to *trust* a terror organisation, even if it goes out of its way to make itself understood? If you decide not to, do you really *prefer* to trust the secret services, since this is your alternative? This is a dilemma. European media therefore often feel caught between the hammer and the anvil – they do not want to be spoonfed by either party but usually have no other option, because informants are hard to find and independent verification is hardly ever possible. Most media therefore choose what they deem to be a possible third way and take to quoting terror experts and news agencies. Nevertheless, most of these have the same problems, after all, and the problem is only perpetuated.

The result is that the above mentioned virtual identity, 'Abu Maysara al-Iraqi', is now referred to as the 'official spokesperson of the Iraqi al-Qaʻida' – even though no one even knows if he is a real person at all. In addition, fake claims of responsibility are frequently quoted in European media because the news agencies seem to have adopted, as their way out of the dilemma, just publishing everything they find on the Internet that looks like al-Qaʿida. At least al-Qaʿida's attempts at disinformation work well – that is safe to say. Thus, the chance to better understand and portray al-Qaʿida, based on its own material, is wasted. The overall result of all these problems is that many European publications do not deal with al-Qaʿida in any analytical way whatsoever, simply because they lack the professional capacities.

Conclusion

While Islamists have always understood the importance of publicity, the use of the Internet is still a new phenomenon. However, we can already see that they are very creative. The Internet is the perfect way to bypass the restrictions they face in the real world and within the borders of their home states. Today, tens of thousands of Islamists log on to the Internet every day and spend considerable time on Islamist and jihadi websites – this clearly shows that we are dealing with an individual sort of 'publicity'.

At the same time mainstream media are reporting more and more about what Islamist online media carry. This is necessary. However, it would be ignorant to deny that this also plays into the hands of the Islamists because it is an indirect way for them to influence an audience they would otherwise not even reach.

Furthermore, it is fair to state that while al-Qaʿida does pretty well in achieving its goals, the European media cannot really be happy with their reporting.

In very important instances al-Qaʿida has even managed to establish something like a dialogue, for example, with the above-mentioned 'peace initiative', after which some European media actually debated the possibility of negotiating with al-Qaʿida. Of course, one must not forget that the prime target was the reaction of the governments, not so much that of the media.

However, overall, while there is no question that al-Qaʿida does cause a reaction in European media, in a more narrow sense there is of course no real dialogue. The European media are clearly on the receiving end, which is in part simply their role – they are not active in this conflict, or at least not supposed to be, not primarily.

It must be noted that we have no feasible ways of instigating real communication with al-Qaʻida-members, and nor do our publications have a lot of impact on al-Qaʻida. True, al-Zarqawi reads the *New York Times*, and Western terror experts sometimes have the doubtful pleasure of seeing one of their articles being discussed by al-Qaʻida sympathisers on the Internet. Nevertheless, that does not change the essential fact that we are mainly listeners. Al-Qaʻida's information and communication strategy has made some things clearer – theoretically. The network cares more than ever about being understood. However, we, or at least the European media, have not yet found a way to deal with this offer in a way that improves our reporting significantly.

Al-Qaʻida, on the other hand, has clearly gained from its strategy. Not only has it closed the gap between its followers and the core organisation considerably, but it has also succeeded in estranging thousands of Arab youths from other news sources by going out of its way to become a news source itself and compete with the established media – al-Qaʻida-TV expressedly aims to counter the 'propaganda' of CNN, Fox News, the BBC and al-Jazeera – even interviewing their own terrorists after attacks and publishing the results to prove that the state-run Arabic and independent Western media lie.

I would like to close with an episode that might be considered minor but in my mind is fairly far-reaching. In 2006 the US military decided to change its policy in regard to publishing the numbers of enemy casualties, a policy that had been followed even during the war in Vietnam. The reason given was that al-Qaʻida's PR must be countered. No other single instance, I believe, proves how influential al-Qaʻida's media work has become – while at the same time highlighting the difficulties Western media have in covering the issue.

Notes

1. Musharbash, Yassin, 'Al-Qa'ida fi Jazeerat al-'Arab: al-Jihad 'ala al Internet', SS., p. 93, in 'Al-Qa'ida fi Jazeerat al-'Arab', *al Mesbar* monthly journal, 2nd edition, al-Mesbar Study and Research Centre, Dubai, 2007. (www.almesbar.net)
2. Yassin Musharbash: 'Die neue al-Qaida. Innenansichten eines lernenden Terrornetzwerks', Köln 2006, p. 114 ff.
3. For a discussion of the real threat that online manuals pose, see Yassin Musharbash, 'Das Internet als Fern-Universität und virtuelles Trainingscamp', online at http://www.bka.de/kriminalwissenschaften/herbsttagung/2007/langfassung_musharbash.pdf. An English summary can be found here: http://www.bka.de/kriminalwissenschaften/herbsttagung/2007/kurzfassung_musharbash_englisch.pdf
4. Musharbash, Dubai Paper.
5. It should be noted that both magazines no longer exist. Other terrorist networks however, some of which have liaison with al-Qa'ida, do publish new magazines based on these examples.
6. Yassin Musharbash, 'Die neue al-Qaida', p. 83 ff.
7. Yassin Musharbash: 'Questions for al-Qa'ida', http://www.spiegel.de/international/world/0,1518,528680,00.html

Contributors

Ehab Bessaiso is a writer, media analyst and a PhD candidate at Cardiff University, Wales. Involved in media studies in the Arab world, he contributed the chapter, 'Al-Jazeera and the War in Afghanistan: A Delivery System or a Mouthpiece' to the book *The Al-Jazeera Phenomenon* (Pluto Press, 2005).

Dr Kai Hafez is Professor and Chair for International and Comparative Communication Studies at the University of Erfurt, Germany, and a senior associate fellow of St Anthony's College, Oxford. He received his master's and doctoral degrees in history and political science from the Universität Hamburg. Most recently, he published *The Myth of Media Globalization* (Polity Press, 2007).

Zahera Harb is a sociologist and lecturer at the School of Modern Languages and Cultures, University of Nottingham. Before that she worked at Cardiff University and as a journalist for Lebanese television. She published recently (with E. Bessaiso) 'British Arab-Muslim audiences and television after September 11', *Journal of Ethnic and Migration Studies*, 32(6) 2006, 1063–1076.

Enrique Klaus is preparing a PhD at the Institut d'Etudes Politiques in Grenoble and is a research fellow at the Centre d'Etudes et de Documentation Economiques, Juridiques et Sociales (CEDEJ) in Cairo. He has published 'The Aîd "Sexual Rage Scandal": Managing the Lack of Information and Orientating a Debate through Categorial Photo-Fit' in Klaus E. & Hassabo C. (dirs), *Chroniques égyptiennes 2006* (CEDEJ, 2007).

Dr Tristan Mattelart is a lecturer in international communication at the Institut Français de Presse (IFP), Université Paris 2 and a researcher at the Centre d'analyse et de recherche interdisciplinaire sur les médias (Carism). He has edited most recently the volume *Médias, migrations et cultures transnationales* (De Boeck, 2007).

Ghania Mouffok is the correspondent for French television station TV5 in Algiers. She contributes to *El Djazaïr* news, *Naqd*, *Le Monde*, *Le Monde Diplomatique*, *La Pensée de Midi* and Radio Canada. She published, among other works, *Etre journaliste en Algérie* (Reporters Sans Frontières, 1996).

Yassin Musharbash is a reporter with Spiegel Online. He holds an MA in Arab studies and political science. His field of expertise is the 'virtual jihad' or the Internet activities of al-Qa'ida and its followers. In August 2006 he published his first book, *The new al-Qa'ida, insights into a learning terror network* (in German, Kiepenheuer & Witsch).

Dr Eric Neveu is Professor of Political Science at the Institut d'Etudes Politiques de Rennes. Among his recent books is *Bourdieu and the journalistic field* (co-edited with Rod Benson, Polity Press, 2005). He has been Dean of the Law faculty in Rennes and is currently Vice Chair of the European Consortium for Political Research and Director of the IEP de Rennes.

Dr Jørgen S. Nielsen is Professor of Islamic Studies at the University of Copenhagen. He holds degrees in Arabic and Middle Eastern studies from the School of Oriental and African Studies, London, and a PhD in Arab history from the American University of Beirut. Recent publications include *Muslim networks and transnational communities in and across Europe*, ed. jointly with S. Allievi (Brill, 2003).

Dr Katharina Nötzold, is a research fellow in Arab Media Policy at the University of Westminster. Specialist in the politics of pan-Arab and Lebanese media, she earned her PhD in communication studies from the University of Erfurt and her master's degree in political science, cultural anthropology and American studies from Leipzig University, Germany.

Contributors

Lawrence Pintak is director of the Kamal Adham Center for Journalism Training and Research at the American University in Cairo and was CBS News Middle East correspondent in the 1980s. He is publisher and co-editor of Arab Media & Society and his latest book is *Reflections in a Bloodshot Lens: America, Islam & the War of Ideas* (Pluto Press, 2006).

Jim Quilty, a former graduate student of Middle Eastern history, is a Canadian who came to Beirut from the US in 1998. Since then, he has written about Middle Eastern politics and cultural production in various English-language publications – most frequently *The Daily Star* newspaper and the specialist magazines *Middle East International* and *Middle East Report*.

Dr Naomi Sakr is Reader in Communication at the Communication and Media Research Institute (CAMRI), University of Westminster, and Director of the CAMRI Arab Media Centre. Her most recent books are *Arab Television Today* (I.B. Tauris, 2007) and the edited collection *Arab Media and Political Renewal: Community, Legitimacy and Public Life* (I. B. Tauris, 2007).

Denis Sieffert is Editing Director of the French weekly *Politis* and author of, most recently, *Comment être (vraiment) républicain* (La Découverte, 2006). He was co-editor in chief at the Agence Centrale de Presse and has taught in different journalism schools, since 2005 at the Institut de Sciences Politiques de Paris.

Maha Taki started her PhD in September 2007 at the University of Westminster in London where the Communication and Media Research Institute awarded her a three-year doctoral scholarship. Her current research explores Internet communication specifically focusing on weblogs in Lebanon, Syria and Jordan.

Husam Tammam is a journalist and was from 2005–2007 a researcher with the Centre d'Etudes et de Documentation Economiques, Juridiques et Sociales (CEDEJ), Cairo. He published most recently *The Transformations of the Muslim Brothers: Dissolution of the Ideology and End of the Organisation* (in Arabic, Madbuli, 2006).

Index

Abbas, Mahmoud 36
'Abbas, Wa'il 255, 256, 258, 259, 260
'Abd al-Fattah, 'Ala' 253, 257
Abdallah, King of Jordan 119, 222, 275
Abdallah, King of Saudi Arabia 123, 140
Abdennour, Ali Yahia 173
al-Abdine Ben Ali, President Zine 123
Abdulhadi, Samir 133
al-Abed, Aref 47, 50, 61
Abou Jahjah, Najla 51
Adie, Kate 52, 60
Agre, Phillip 190
Ahamd, Aït 179
Alexandre, Laurien 157–8
Almond, Gabriel 246
Alterman, J B 109
Alwaleed Bin Talal, Prince 270
Amer, Jamal 123
Amir, Karim 254
Aoun, Elie 137
Aoun, General Michel 270
al-Arabaya (Arabic satellite TV) 21, 98, 101, 123, 136, 137
Arendt, Hannah 234
al-Aris, Safi 47
Atallah, Samir 134
Attieh, Rowayda 275
al-Awa, Dr Nuhammad Selim 20
Awada, W 59
Ayntrazi, Tarek 138

Barakat, Ali 203

Barnett, Sam 137
Bartlett, Dan 77
Bassets, Lluis 158
Bastardes, Enric 158
Battah, Habib 279
Bayart, Jean-François 224
Bayat, Asef 224
Beelman, Maud 78
Bell, Martin 52, 60, 72, 74
Ben Ami, Shlomo 36
Benjedid, President Chadli 173, 179
Berri, Nabih 55, 270
Bessaiso, Ehab 71–97, 297
Bin Laden, Osama 40, 42, 235, 239, 286, 289, 290
al-Bissar, Salma 148
Blair, Prime Minister Tony 83, 238
Blumer, Herbert 107
Booz Allen Hamilton 139
Boucher, Richard 77
Boulos, Paul 138
Bourdieu, Pierre 186, 221, 225
Boyd, Douglas
  Broadcasting in the Arab World 157, 158
  Straubhaar, Joseph D and Lent, John A, *Videocasette Recorders in the Third World* 159
Boyd-Barrett, Oliver 134
Brown, Ben 81
Brown, Robin 72
Browne, Donald R 156

Bush, President George W 37, 38, 43, 71, 75, 82, 83, 120, 233, 238, 244
Byrne, Ciar 86
Bytwerk, Randall 77

Calabrese, Andrew 82, 83
Cardiff School of Journalism 85–6, 87, 88
Champagne, Patrick 222
Chaoul, Melhem 271
Chidiac, May 118
Choueiri, Antoine 138
Clarke, Victoria 77
Collovald, Annie 215
Colomb, Dominique, 166–7
Coméliau, Christian 167
Cossery, Albert 224
Costandi, Michel 141

The Daily Star (Lebanese newspaper) 147–51
Danish People's Party 22, 25, 26, 27, 29, 31, 32
al-Deeb, Ameen 202
Defrancis, Suzy 77
Desrosières, Alan 220
Diène, Doudou 21
Djaout, Tahar 174
Downing, John D 160
Dray, Joss 35
Durham, F and Singer, J 53
Durkheim, Émile 213

Eilders, Christiane 80
Elias, Norbert 213
Elleman-Jensen, Uffe 26–7
Enderlin, Charles 38
Eskew, Tucker 77
Espersen, Søren 31

Fandy, Mamoun 186
Farge, Arlette 222
Fatany, Samar 122
Ferjani, Riadh 163
Fisk, Robert, *The Great War for Civilization* 174

Fleischer, Ari 76
Foucault, Michel 190
Fox News 76, 80, 84, 140, 141, 235
Frevert, Louise 17, 30
Future TV (Beirut TV) 61–2, 270–5*passim*

Gabr, Karam 262
Gerschenkron, Alexander 226
Gibson, John 76
Grosser, Alfred 240

Habermas, Jurgen 212, 217, 264, 268
Hafez, Dr Kai 229–51, 297
El Hage, Nakhle 136
Hale, Julian, *Radio Power* 157
Hall, Stuart 102
Hallin, Daniel C 103–4
Hamad Ben Khalifa al-Thani, Sheikh 106
Hamrouche, Mouloud 175, 176
Haniyeh, Ismael 42
Harb, Zahera 45–70, 297
Hariri, Prime Minister Rafiq 55, 61, 123, 223, 270
Haroub, Khaled 269
Has, Amira 224
Hasan, Manal 253
Hashem, Roy 270
Hassan, Ammar 277
Haykal, Muhammad Hassanein 119, 258
Helal, Ibrahim 88
Herbst, Susan 212
Herman, Edward S 165–6
Hindawi, Ahmad 58
Hine, Christine 187
Hizbullah 40, 45, 54, 108, 124, 270, 285
Holm, Ulla 31
Hotelling's Law 135, 142
Huntington, Samuel 230, 243
Hussein, Saddam 76, 82, 83, 101, 104, 110, 223, 245
  toppling of statue 87, 88, 89, 121

al-Ibrahim, Waleed 140
al-Iraqi, 'Abu Maysara 291, 292

Index

Iskandar, A and El-Nawawy, M 56–7
Det Islamiske Trossamfund 18, 19, 26

Jalloul, Mahmoud 47
Jammal, Christine 279
al-Jazeera (Qatari TV channel) 11, 76, 88, 89–111*passim*, 117, 133–4, 137, 164, 235, 239, 245–6, 258, 262
Jubair, Adel 136
Juste, Claus 20
Jyllands-Posten (Danish newspaper) 17, 18, 20, 21, 27, 29, 30–1

Kamel, Sheikh Saleh 133, 277
Kassir, Samir 46, 118
Katovsky, Bill 74, 75
Kellner, Douglas 79, 83
Kenneth, Colonel 73
Kepel, Gilles 225
Khader, Samir 121
Khalil, Joe 273
Khalil, Magdi 268
Khanfar, Wadah 133–4
Khatami, Ayatollah 247
Khayyat, Tahseen 270
Khomeini, Ayatollah 247
Khouri, Rami 135
Kitch, C 51
Klaus, Enrique 252–67, 297
Koch, Professor Henning 29
Kouyoumdjian, Zaven 57, 59–60, 272
Krarup, Søren 25
Krichen, M'hamed 121, 125
Kumar, Deepa 79, 80, 81, 86

Lamloum, Olfa 98–115, 164, 225
Langballe, Jesper 25
Le Bon, Gustave 224
Lee, Martin A 79
Lewis, Justin 73, 76
Lockwood, Group Captain Al 86
Lydersen, Kari 75, 86
Lynch, Marc 246, 268

McChesney, Robert W 165–6
McClure, Laura 82

Madani, Lotfi 163
al-Mahariq, Sameh 119
Mahfouz, Naguib 224, 256
Mahmoudi, Abderrrahmane, 181
Mahmud, 'Abd al-Min'im 261, 263
Majzoub, Nadine 47, 52
Makari, Farid 53
Mannion, David 60
Mansour, Ahmed 120
Maroit, Nicholas 223
al-Mashnouk, Nouhad 62
Mattelart, Dr Tristan 155–71, 298
  Le cheval de Troie audiovisuel 160, 162, 166
Mehri, Abdelhamid 178
el-Menawy, Abd el-Latif 125
Middle East International (magazine) 147, 151–3
Mikkelsen, Brian 17
Miller, Danny and Slater, Don 186, 187
Miller, David 73–4, 83
Miller Laura 86, 87
Møller, Per Stig 20, 21
Mouffok, Ghania 172–83, 298
  Etre journaliste en Algerie 174
Moukaled, Diana 61, 62
Mroué, Jamil 148
Mrowa, Kamel 148
Mubarak, President Hosni 119, 196, 252
Muhammad, Mustafa 203
Muir, Jim 151
Musa, 'Amr 20
Musharbash, Yassin 285–95, 298
Muslim Brotherhood 22, 41, 124, 195–208*passim*, 254, 259, 261

Naim, Fouad 47, 47–8, 50, 56, 61, 63
Najem, Dr 57
Nasrallah, Hassan 124
Nasser, President Gamal Abdul 119, 124, 134, 148, 157
Negt, Oskar 264
Neveu, Dr Erik 211–28, 298
Nielsen, Dr Jørgen S 17–34, 298

Nordenstreng, Kaarle and Schiller, Herbert I 156, 159
Norris, Pippa 231
Nötzold, Dr Katharina 268–84, 298
Nunberg, Geoffrey 85

Orwell, George 212
Ouyahia, Ahmed 172
Ozouf, Mona 212

Pantti, Mervi 51
Pintak, Lawrence 116–28, 235, 299
Powell, Colin 82–3
al-Prince, Hassan 203

al-Qaʿida 40, 82, 83, 105, 108, 285–94 *passim*
al-Qaradawi, Dr Yussuf 20, 107, 275
Quilty, Jim 147–54, 299

Rahbani, Elias 274, 275
Ramadan, Tariq 40, 41
Rampton, Sheldon and Stauber, John 83–4, 86, 87–8
Rasmussen, Prime Minister Anders Fogh 18, 20, 20–1, 27
Rodinson, Maxime 35
Rose, Flemming 17
Rosecrance, Richard 236
Roy, Olivier 107, 108
Rushdie, Salman, *The Satanic Verses* 22, 24, 217
Rutherford, Paul 78, 81, 84–5

Said, Edward, *Orientalism* 247
Sainath, P 84
Sakr, Dr Naomi 103, 131–46, 163, 240, 299
Sambrook, Richard 86
Schechter, Danny 77, 79, 80, 85
Schifferes, Steve 82
Schudson, Michael 119
Scott, James C 161–2, 224
Seale, Patrick 151
Seib, Philip 74, 110
Semelin, Jacques 159, 160
al-Sharqawy, Muhammad 257, 260

Sherman, Steve 151–2, 153
Shmait, Walid 81
Sieffert, Denis 35–44, 299
Simpson, John 76
al-Sini, Othman 116, 125
Siniora, Fouad 270
Slevin, James 187
Snow, Nancy 78, 84
Solomon, Norman 79
Sreberny, Annabelle and Mohammadi, Ali *Small Media, Big Revolution* 161, 162–3, 165

Tabbara, Riyad 62
Taki, Maha 184–94, 299
Tammam, Husam 195–208, 299
Tatham, Steven 77
Taverner, Angus 77
Télé Liban (Lebanese TV) 46–7, 49–64 *passim*, 270
Tilly, Charles 223
Tomaselli, Keyan and Ruth 166
Tueni, Gibran 118
Tumber, Howard and Palmer, J 74, 76, 77

Umar, Mullah 241
Usher, Graham 151

van Gogh, Theo 29
Vittin, Theophile 161

Walters, Dennis 152–3
Webster, Frank 74
Wehbi, Haifa 276
Wheeler, Deborah 185, 187
Whitman, Bryan 74–5
Wieten, Jan 51
Winters and Giffin 74

Zaret, David 217
al-Zarqawi, Abu Musab 291, 294
al-Zawahiri, Aiman 291
Zayani, Mohamed 110
Zein, Melhem 275
Zilo, Alexander 133